AF278939

The Planets
in Dr Steiner's
Agriculture Course

An Answer to the Question

ISBN 978-0-9517890-8-7

Contents

The Problem

The impetus for providing a hardcopy version of this topic comes from the circular letter No 119, from the Agricultural Section of the Anthroposophical Society of Switzerland , which details their efforts to produce the 9th edition of Rudolf Steiner's *Agriculture Course* in time for the 100th anniversary of the original lecture cycle. In their notes about their changes there is this passage:

Direct and indirect working of the Planets

There is a contradiction between lecture 2 and lecture 6, which the editors of the new edition draw attention to, without being able to finally resolve the problems involved. In lecture 2 it is said the moon and the inner planets, Mercury and Venus, work together with the sun directly on the plants above ground, the outer planets, Mars, Jupiter and Saturn, on the other hand, work with the sun in a roundabout way via the earth. In lecture 6 it is exactly the other way around: the outer planets work directly on the plant above ground and the moon works with the inner planets, Mercury and Venus, indirectly in a roundabout way via the soil. Important statements about the effect of the moon and the planets on the plants are not impaired by this contradiction!

After 100 years this contradiction is still unresolved …. by them.

I however, first put forward a solution to this problem in the 1980s. I revisited this topic in 2019 while investigating Rudolf Steiner's comments about Plant Growth. Since 2000, I have been to the annual Biodynamic conference in Dornach 3 times - traveling to the other side of the Earth - and given 5 hours of presentation in which I provided this solution. My efforts appear to have fallen upon deaf ears. I have had no communication with anyone from 'the centre' about my suggestions, even though my answer is completely consistent with the Anthroposophical medical worldview. This possibly is because I am not a member of 'the club', as outlined by RS in the Lesson on the Eighth April 1924. Which is a very sad indictment for true seekers of truth. What I am suggesting is placed within a larger context than European BD appears to be allowing themselves. This larger context - the medical worldview - allows for more complex perspectives to be bought to bear upon this subject.

From what I can tell, the Euro BD story is based predominately upon a 2-fold polar image of a push from above downwards, meeting a push from below upwards. Calcium and Silica mediate these processes, but which is Calcium and which is Silica?

This basic polarity is further enhanced as manifesting through the primary formative influences called the four Ethers. These also work as a polarity of Warmth and Light against Water/Chemical and Earth/Life. This Etheric activity is seen as the dominant Formative Force upon Lifeforms. This Physical Etheric organism is then organised further by the Astral, as a outside imprinting force. The Spirit is not often present as an active player other than via the Human. This worldview was very ably conveyed by Richard Thornton Smith in his 2009 book 'Cosmos Earth and Nutrition' . When I asked him why he presented this worldview, he told me he attempted to capture the 'accepted' biodynamic story of the time in Europe.

This worldview does not allow for a fourfold polarised planetary reference, as given in the Agriculture Course, and enlarged by Lievegoed, to make sense.

The problem with the Ethers

A significant problem with Dornach's story, being based around the Ethers, is that **the Agriculture Course does not tell the story of the Ethers**. There is only one undisputed reference to the Ethers in the whole course.

The course tells two different stories of how natures energies organise. Most importantly Dr Steiner talks of the four primary Astronomical Spheres, - Galaxy, Planets, Atmosphere and Earth - directing their energetic activities under the names of Spirit Astral Etheric and Physical, into Substance. We meet them firstly in Lecture 3, which tells us how the main energetic activities moves into Chemistry, while Lectures 4 and 5 gives us preparations to control them and lecture 7 tells us how they work in the landscape.

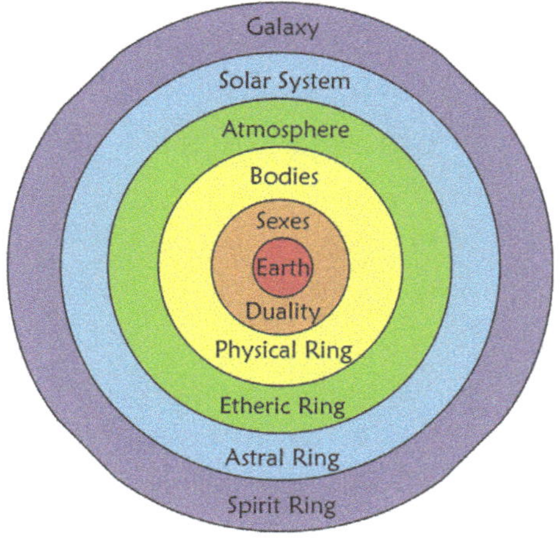

1

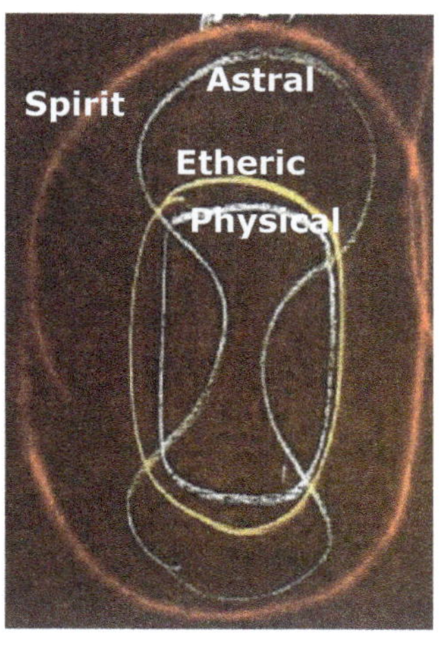

2   RS 1924 Pastoral Medicine

1 is our external environment

2 is how RS drew the bodies inside a Human.

The second story is told firstly in Lecture 2, where we are told of how these primary activities manifest within the Physical sphere, as Physical Formative Forces, (PFF) called the Cosmic Forces and Cosmic Substance, and the Earthly Forces and Earthly Substance activities. These work through the 3 fold Physical Organism, using the substances Clay, Sand, Humus, Lime. The best picture for this is from lecture 8.

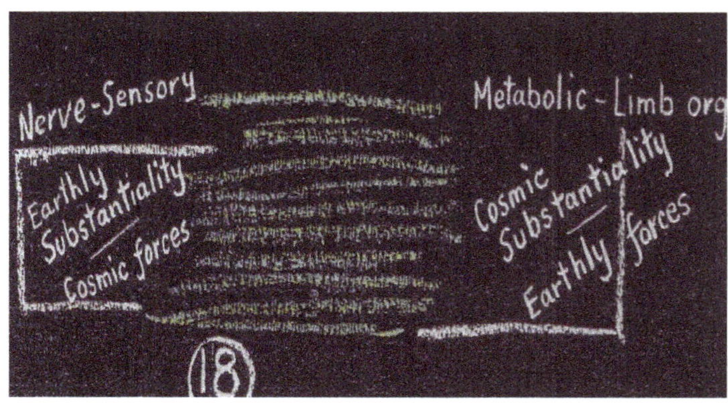

3   RS lec 8 Ag C 1924

This is our first introduction to the second theme of the Cosmic and Earthly Activities, which is enlarged upon further in the 4th 6th and 8th lectures. Each lecture tells us of a fourfold story in a different sphere of activity. The polaric relationships of the Cosmic and Earthly activities are different to the way the Ethers organise. So there is no easy fit for the Ethers story, into the stories told in the Agriculture Course.

Adding to the confusion is that the Ethers are not THE primary formative influences. They are the players you might see, but they are just following orders from higher up. If we use a building site analogy. The Ethers, which are found in the Earth Atmosphere, (12) are the on site builders. They take their direction from the Master Builder who are the Planets, who in turn takes their direction from the architect, who is the Stars. If we want to know and change the plan we need to talk with the architect.

Dr Lievegoed's introduction gives us the energetic context we have to work with.

" In every living organism
Physical, Ethereal, Astral and Spiritual forces are active.
In the plant, the *threefold* Physical and
the *fourfold* Ethereal forces, work from within outwards,
the Astral forces work around it,
and the Spiritual forces from the starry distance,
where the archetypes of the plants are formed"

"The effects of the Etheric Formative Forces are always general ones.
Never does an organism come into existence merely through the Ethereal processes,
but only when an Astral principle presses its seal upon the Ethereal"

" The working of the Astral principle is carried into this Physical-Ethereal,
as a streaming, moving activity. Manifesting archetypally in a sevenfoldness"

"The Spiritually working principles gather together the form
into a specific species and are archetypally arranged
according to the 12 principles, revealing themselves cosmically
in the Zodiac. It is only through this spiritual element that
'the plant' becomes a rose or a sage,
'the animal', a lion, and 'man a definite individuality."

"In the plant the seed is the carrier, or point of attachment,
of the Spiritual forces *which* form the species.
But the seed only unfolds when surrounded by
Physical, Ethereal and Astral elements in such a way
that these forces can stream freely together."

Here the Ethers are put in their rightful place. The general BD thinking has to move from the Ethers focus , in both directions. Outwardly, by seeing the Etheric in its place amongst all the Bodies, and Inwardly by seeing the Cosmic and Earthly activities (PFF) as a different story, occurring within the Physical Body of things. This is a move inwards into the Physical. The Etheric Forces and the Physical Forces are both fourfold, however the way they work internally are different. The PFF polarise as micro polarities. While the Ethers are generally talked of within the macro polarities (pg 18) of the Cosmic Ethers of Warmth and Light polarised against the Earthly Ethers of Chemical/Water and Life / Physical. The Physical Formative activities, use the micro polarities and have the Warmth related Cosmic Forces polarised against the Physical Life Ether related Earthly Substance, while the Light related Cosmic Substance is polarised with the Chemical / Water ether related Earthly Forces. They are two different stories, and why Dr Steiner uses different words to tell their stories.

The Ethers story is told in Man as Symphony of the Creative Word - Lecture 7, 2nd November, 1923   http://wn.rsarchive.org/Lectures/Dates/19231102p01.html

How the Ethers story has become the dominant reference for Biodynamics is a separate mystery to solve. (13)

Once we appreciate these inner structures of the Course, we can understand some of the deeper offerings of the course.

But first lets look at the planets problem.

In Lecture 2 we have this
"The forces at work above the earth are immediately dependent upon what we will regard for the time being localised on the planets Moon, Mercury and Venus. These planets in strengthening and modifying the effects of the Sun, exercise their influence on all that is above the earth surface, while the more distant planets lying outside the earth's path round the Sun strengthen and modify the effects of the solar influences which penetrate upwards through the earth. Thus the growth of plants is affected by distant heavens in so far as it takes place underground, and by the nearer heavens in so far as it takes place above ground; and the influences upon vegetable growth coming from the expanses of the Cosmos do not shine directly down upon the earth but are first absorbed by the earth which then causes them to radiate upwards. What come from beneath as good or bad vegetable growth are really the cosmic influences which are reflected from below; whereas in the air and water above the earth the Cosmos exercises its power directly. The direct cosmic in-streaming is stored up beneath the earths surface, and from there it works back. The inherent qualities of the soil affecting the growth of plants are dependent upon these stored up influences. (Later we shall consider the case of the animals). The soil still retains in it the effects of influences dependent upon the most remote parts of the Cosmos, which need to be considered in connection with the Earth. These effects are found in what we know generally as sand and rock; the substances which do not absorb water, which are ordinarily supposed to contain no nutritive elements whatsoever and which nevertheless play a every important part in the promotion of growth. These minerals are entirely dependent upon the activities of forces coming from the remotest parts of the Cosmos….. …Now the effects which have been brought about in the root through silicon (cosmic outer planets) must be borne upwards through the plant. It must stream upwards and there must be a constant interaction between the cosmic forces that have entered into the plant through silicon and those that are active above - forgive me in the `belly" and

that supply the "head" below with what it required. True the " head" must be provided for out of the Cosmos but this process must interact with that which takes place above ground in the "belly". The forces coming in from the cosmos and being caught up underground must be able to flow upwards again, and the substance which brings this about is clay. Clay is the mediator through which the cosmic activity in the soil is enabled to work from below upwards. In actual practice this will give us the key to the handling of both clay soil and sandy soil according to the particular way we may wish to cultivate. But we must first know what is actually happening. How clay is to be described and treated in order to make it fertile are important but secondary considerations. The first and foremost thing to know about clay is that it promotes the cosmic upward flow. "

In Lecture 6 we have this

"It was pointed out that we must learn to distinguish those forces which arise in the cosmos but are absorbed by the earth and work upon plant-growth from within the earth. These forces coming from Mercury, Venus and Moon and act not directly, but through the mediation of the earth. They must be taken into account if we wish to follow up how the mother-plant gives rise to a daughter-plant, and so on. On the other hand, we have to consider the forces taken by the plant from the outer-earthly, and brought to it by way of the atmosphere from the outer planets."

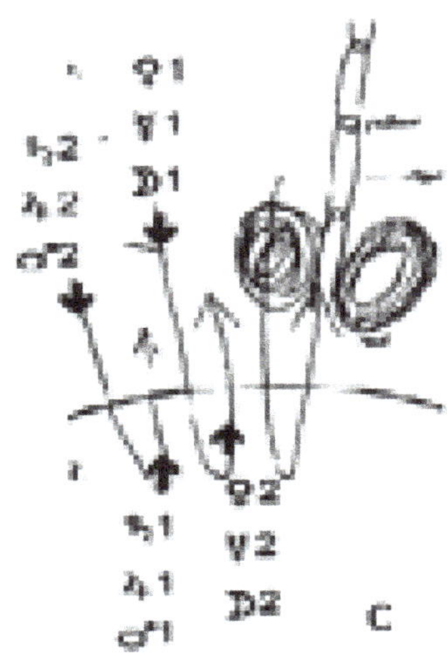

☽	Moon	♅	Uranus
☿	Mercury	♆	Neptune
♀	Venus	♇	Pluto
♂	Mars	Persephone
♃	Jupiter	Vulcan
♄	Saturn	☉	Sun

"Now I want you to imagine that Diagram C represents the earth level, where the influences of Venus, Mercury and Moon I **enter into the earth and stream again from below upwards**. These are the forces which cause the plant to grow during the season, later produce the seed, and by means of this seed a new plant', a second plant, then yet a third and so on. (I indicate this schematically). All this goes into the power of reproduction and streams on into the succeeding generations. The forces, however, which take the other path, remaining above the earth level, come from the distant planets. I can draw this schematically in this way. These forces cause the plant either to spread into its surroundings or to become fat and juicy, to build matter into itself such as we can use for food because it is produced again and again in a continuous stream." "From this we are able to see how we must proceed if we are to influence plant-growth in one way or another. We have to take account of these two sets of forces."

Diagram C, with my additions, seems to answer the problem adequately, however Dornach can not see this.

To address this planetary contradiction ,we need a further step to be added to the single '7 Planets' story, most of Bidoynamics works with. Dr Lievegoed in his 1951 lectures (4) provides us with the entrance point. He tells us that all the 6 planets have two actions, one as they come towards the Earth, he calls Incarnating and another as

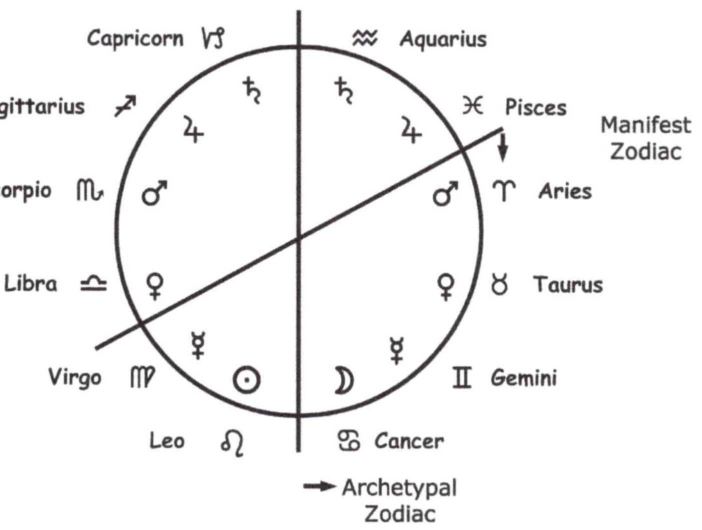

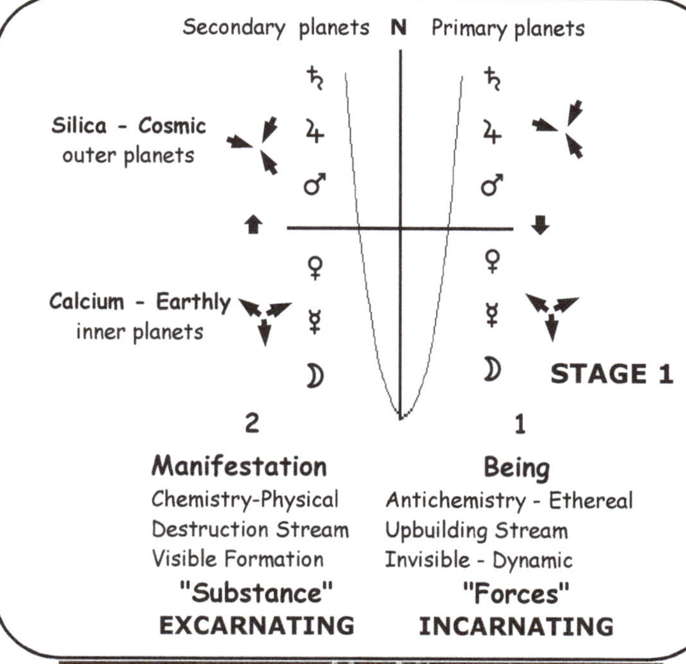

they leave the Earth called Excarnating. This planetary order is present in the Zodiac. He is bringing Steiner medical insights to our aid. The medical picture is somewhat more complex than a simple polarity image, and asks us to look for the four primary activities at every level of the 3 fold physical manifestation. Initially though it is enough to look for two dominant players in each physical system. (pic 3) One drives the expansive processes and the other drives the contractive processes, in each physical system. This is also the story of the Cosmic and Earthly processes in the Agriculture Course. The trick though is how we get from Dr Lievegoed to the Agriculture Course, and onto the Seasonal pictures.

We can not escape that Dr Steiners stories are complex, and multi layered with many moving parts. He often uses completely different languages to tell the same story. The planets however are often the only common reference between the stories. So the Planets story are the key players to follow through the various stages. They keep their name throughout this interactive phenomena.

We are playing a **6 layered game of chess, that moves through three different stages.** Dr Steiner describes this without telling us clearly where he is in this complex. He simply describes what he sees, from where he is, without giving us the xyz coordinates. All the Layers are energetically linked, and all the individual parts can exert their own force, to influence the larger

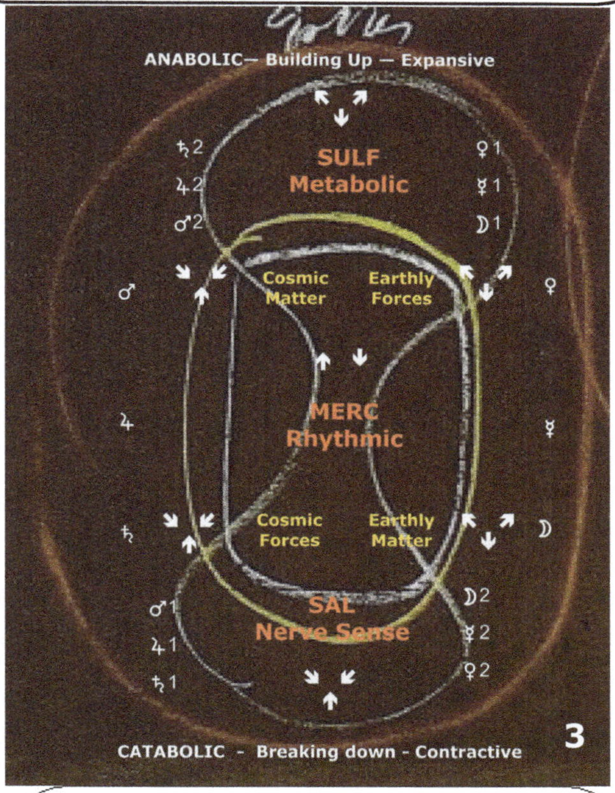

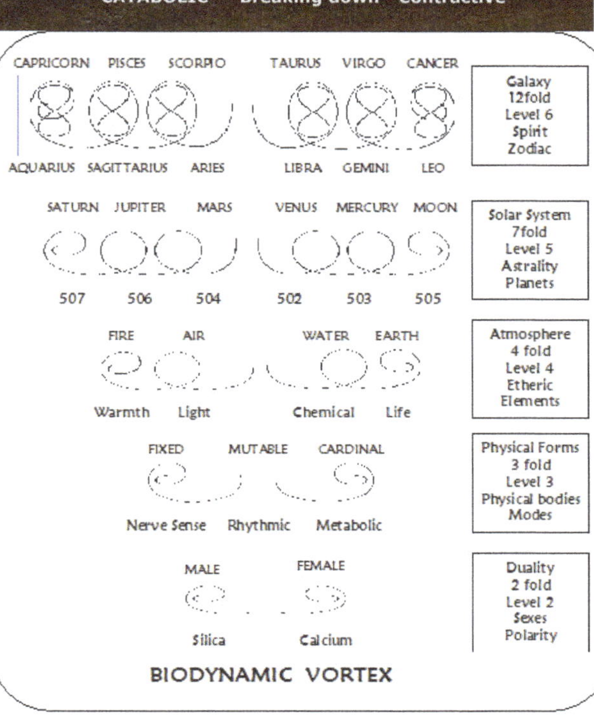

energy vortex. Creation is a very dynamic Being, and keeping track of it all is a trick. It is so complex I have needed to exteriorize the stories into Pictures. Dr Steiner's stories are based upon Astronomical realities, of various kinds. Simply put he has observed the basic universal order of 1 unity, the 2 polarity ,3 Physical Modes ,4 Elements ,7 Planets, and the 12 Constellations. So any Pictures derived from following these Universal Laws can provide us with potentially truthful information, as it is based upon real astronomical truths. I rarely know what my pictures mean when they are first created, as I have just followed the 'rules', but by looking at them further many potential truths have arisen.

Dr Steiner gives us signposts along the way. He provides a jigsaw puzzle of ideas, images and applications. We have to be able to fill in the gaps between the signposts, and organise the parts into a coherent story that does exist amongst it all. This is a very Neptunian activity. As we would expect from RS birthchart , (14) where we see a lot of Neptunes influence. He forces us to merge an array of diverse imaginations together. 95% of the leading BD people, whose birthcharts I have seen, have a strong Neptune active. Interestingly several have a Sun square Neptune, which suggests they can be a tad eschew with their imaginations.

We are lucky to have 1000s of medical doctors working on this same worldview. They have refined Dr Steiner's story into a credible real world modality. All we have to do is find the connecting points to the Agriculture Course, and all their science is available to us. Then we can ask what are we going to do with this insight. How do the Agriculture Course passages read now.

Dr Steiner's story of how nature works is bigger than the Ag Course. This is only one part of the overall story. We have to include Dr Bernard Lievegoed's 1951 book on the Planets, along with two 1923 lectures, where the Ethers are mentioned, to get an fair overview of RS suggestions, towards Nature. We have to understand where RS comes from (Lievegoed) and where he goes to in these 1923 lectures. This is where we find three distinct stages of development, and three different planetary patterns. Nothing is 'wrong' , we just have to see where everything is 'right'.

I am the Periphery. I am not German, I am not an Anthroposophist, I live at one of the farthest Eastern places on Earth. The first sunshine of each new day, hits the hill behind my house. This is the farthest point away from Europe. I am one person, with a garden, a keyboard, and a successful BD business that has funded my interest in Dr Steiner's suggestions, for 30 years. I am also an avid Astrologer and this brings many useful understandings of this whole story, to the conversation. I have done my own study of Dr Steiner with very little outside input, and I have come to my own conclusions of what he is saying. I am so glad that at the end of the rabbithole, the story is the same story told in the medical lectures. Whew, its not MY story. How I make sense of it all is my story, but what we are making sense of has many friends, who tell a good tale. All this has allowed me to look at the information through clear eyes, free of the existing Anthroposophical and Biodynamic beliefs, which are acknowledged as inadequate.

This is what I see.

My Process

Through his Goethean Science Dr Steiner has given us a way to generate new information. With his use of 'universal laws' based upon the astronomical reality of the Universe, we can use the observable order we find when looking at the universe, as the 'universal' reference for our observations of Manifestation.

> "... when we want to understand the plant, we must bring into question not only plant animal and human life, but the whole universe. For life comes from the whole universe not only the Earth. Nature is a unity and her forces are at work from all sides. He who can keep his mind open to the manifest workings of these forces will understand her. " R.S. Pg 70 1938 Agriculture.

The external organisation in Galaxy, Solar Systems and Planets and their Atmospheres, are fundamental organised realities, real energetic dimensions within the Creative Vortex. Physics and the Golden Mean maths that describes their form are the flesh upon these bones. Dr Steiner describes these patterns resonating into the forms of Life, as boldly as in the four Kingdoms of Nature. (21 page 18.) Where each kingdom embodies a further astronomical sphere. The mineral kingdom is just physical Earth. The Plant kingdom embodies the life giving Oxygen filled Atmospheric activities, while the Animal Kingdom adds the sensitive movement we find in the planets of the Solar System , while the Human Kingdom embodies the individualising forces of the Stars.

Dr Steiner used these resonating patterns in many of his natural science lectures. We can see a structure long used by Astrology, and RS talked of as the 12-fold, 7-fold 4-fold 3-fold 2-fold and unity, order. My 'Bernie and Glen' piece adds to this conversation. (22) We can also remember RS suggestion at the end of lecture 5 in Cosmic Workings in Earth and Man — *"And one must also know astrology, the science of the stars, if one wishes to understand the formation of cambium"*.

My observation method, (as I can not really says it is identical to RS's) asks us to collect all the available scientific information on a topic, then look at this information. Look for patterns in the scientific information in front of you, identifying any of these numbers. Once the Number Dimension is established for your cluster of information, we can look for further organisation, within it. In my chemistry investigation I observe three groups of elements, each with a different number. There are 8 groups of Major elements, 10 groups of Transition elements and 14 groups of Lanthanide elements. This suggests the numbers 4,5,and 7 can be organising principles for these elements, and

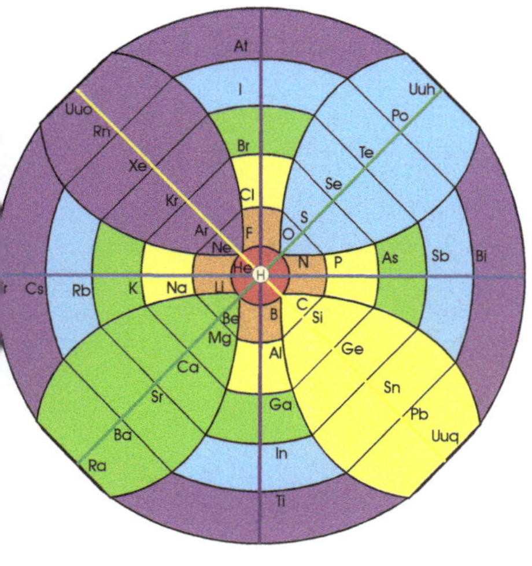
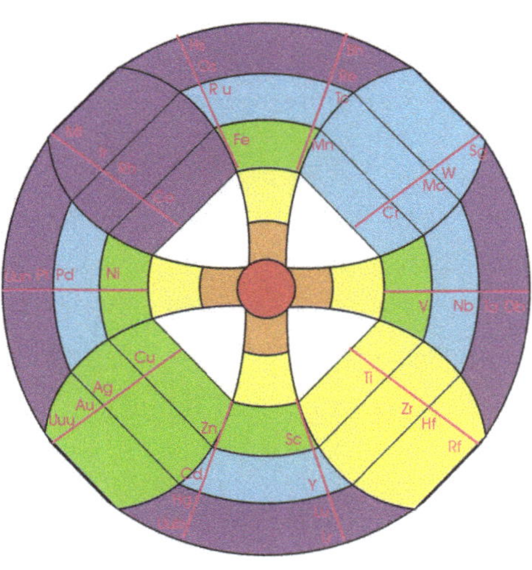

that there may be a inner polarity between the one group of 7 elements and the other group of 7 elements, within the 14 Lanthanides and so on. I then used melting points, valances, atomic weight and the elements uses, to further define an order that can be further related any other information I already have related to the number 7, within a circle. (1)

In this way we can build up reference points for establishing possible truths that can be further researched. We can ask how does the inner order enliven the information and what insights have been generated through that exploration?

This process has led to a large Astrological Equation, that moves according to the context it finds itself in. The context always has some relationship to an Astronomical basis, so has a basis connected to truth. However the Universe is a big place and there are many context changes. Keeping an eye on the astronomical reality is the shining star in the night. The Astrological rules provide some boundaries to form our findings into things worth investigating. Astrology is based upon the Physics of Creation.

This new information should be considered speculative, however the speculation is based on a proven theorem. So it has a very high potential of being right, and is thus worth pursuing further. We must look for ways to prove this speculation, or not. Keep doing this.

I see this activity as Dr Steiner's first stage of Perception. He then describes a second stage as what happens when our intense study and observations reach a peak, which in my case is usually accompanied by some frustration. At this point I release or give up on the thought form, and into this space flows new possible ways of approaching the subject. An answer to the question that was unable to answered before, often appears.

We can imagine this process as our Internal Spirit , through focus thought , enlarging in size and pushing out into the surrounding environment. Our 'environment' is initially created by the real energy coming from ALL the billions of Stars. This energetic holographic field is called the World Spirit sphere. It is within this energetic field that the planetary Astral and atmospheric Etheric influences can do their further organisation. When we actively think and imagine thought forms we are pushing out into the World Spirit sphere. We are consciously looking for inspiration for our questions, pushing in the Collective Unconscious, held within the field, with our meditations on our subject. Most people can receive some direct help from this activity. However when we release our focus and the push out wards, the World Spirit and Collective Unconscious push back and flood into the empty space in our consciousness, left by our thought form. The answer is supplied out of the Universal reasonance responding to our thought development. Thought is indeed a formative force and the universe responds. Everything is in movement and in response to everything else.

The third stage of RS perceptive process is when we can meet the energetic individuals in the surrounding field, who are focusing these Cosmic Thoughts back to us. Here personal conversations with these energetic individuals can be had. These Beings can be of all manner of nature, from dead people through to Elemental, Planetary and Star entities. Whoever you wish to communicate with is usually available. Just ask. Receiving what they have to say is often my difficulty.

This is the process of 'knowing all things'. Every thought and feeling ever had by anything on Earth, leaves an impression within the energetic field of the Earth. If we can become conscious of how we are embedded within this field, we can access Inspirations, Imaginations and Intuitions. Ask and the field replies.

When this process become fully conscious one is 'Enlightened', and is one step beyond Pluto and at the sphere of Persephone, which Dr Steiner calls 'Solid Land' in lecture 3 of Occult Science. Creative visualisation or directed intentions leading to manifestation, becomes almost instantaneous, in this sphere.

During all these stages we are generating really good new information.
Dr Steiner suggested that his followers join together and listen to each others insights as 'spaciously' as possible. His path is about discovering new truths which work with the Universe, to produce more ordered and sustainable Life.

This is a new frontier and we have the tools to explore it consciously and safely.

The 7 Planets

Dr Steiner's references to the planets within the Agriculture Course are scant, quite confusing and thus easily neglected. There are two ways of approaching them. One is quite simple, while the other is somewhat more complex. We best start with the simple approach.

In the first lecture we are given the external 'World' big picture, of our planet. Dr Steiner is describing the primary polarity between the contractive processes, we suspect are coming from above as Cosmic influences, working through the outer planets, and through Silica into the Earth. These activities are polarised against the expansive Earthly processes, working through Calcium and the Inner Planets. This image reflects the creation myths of most cultures, where the Sky Father meets with the Earth Mother and the children or Life are created in between the two.

So while, Dr Steiner does not say clearly, which set of planets are coming from which direction, it is fair for these relationships to be established, for his commets. The planetary references here suggest the 'Earthly' inner planets come from below, supported by the Earth Mother, while the Outer Planets come from above, supporting the Sky Father. I suspect most people leave the first lecture with this image.

Elsewhere RS talks of the Human having a similar organisation. The outer planets work in their head, while the inner planets work below in the metabolic system.

But then in the second lecture we have something different, where we are given the internal organisation of the plant based farm organism. In this story the Inner planets are coming from above, and the Outer planets are coming from the Earth. The plant is to be seen as the opposite of the Human. The contractive 'head' of the plant is the roots, while the reproductive parts of the plant are above the soil. So in their physical organisation it is fair to outline the planets in this manner for plants.

For animals we find in the latter section of lecture 2, RS shows how the animals, while orientated on the horizontal plane, has a similar organisation , that the head is ruled by the Sun and the outer planets , while the back of the animal is ruled by the Moon and the inner planets.

This planetary order is interesting. It is based upon the **lengths of the various planets cycles around the Sun**, and in the case of the **Moon around the Earth**. So these are **real rhythms within the electro magnetic body of our environment**.

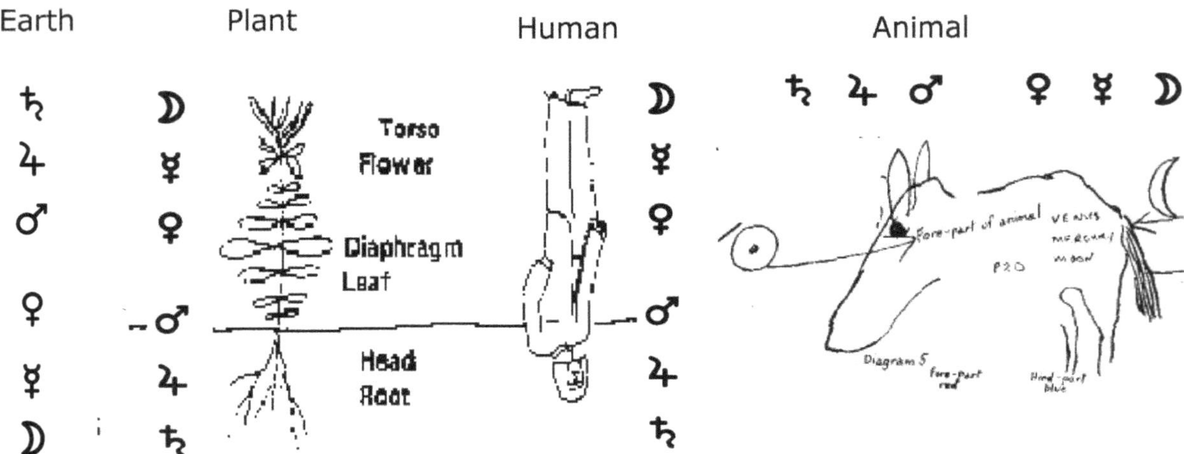

This same order found in the planets relationship to the Zodiac. (pg 23) This is an interesting mix of Heliocentric and Geocentric realities, as we experience them on Earth. The inner planets are not our Geocentric experience. The Sun is seen from the Earth, yet it is still given its Heliocentric Star status, and is seen as the Spirit providing a central balancing and directing influence of the other planets.

With RS's use of this order we are presented with an image of the 'earthly' inner planets - those closer to us than the Sun, including the moon - and the 'cosmic' outer planets, those more distant than the Sun, and as far as Saturn.

To observe Reality we need to keep both Heliocentric and Geocentric 'realities' in mind. Reality says the planets are within concentric spheres of magnetic rings around the Sun.

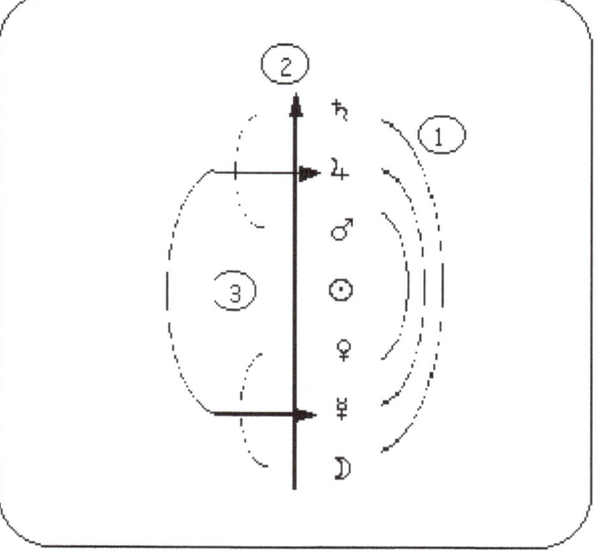

However, seen from the Earth they start at the ionosphere, and extend to the edge of our solar system. At present this pattern extends out to the orbit of the most recently discovered planet, Pluto. However, when dealing with manifest life it is common to use only the 'traditional' heavenly bodies that we can see with our eyes: Sun, Moon, Mercury, Venus, Mars, Jupiter and Saturn. The three further planets - Uranus, Neptune and Pluto - were consciously discovered in more recent history and are associated with areas of extra-sensory perception and the collective unconsciousness. The 'original' seven planets, are connected to the physical manifesting streams of the universe. Saturn is seen as the limit or boundary-giver.

We experience the planets from Earth, so we have to also consider their apparent retrograde movements as a geocentric phenomena, even though this does not occur heliocentrically. Given our central position in the planetary order, it so happens that we have three planets on each side of us. This is just as it is. Within Astrology the planets can be organised into several patterns of relationships, however Dr Steiner uses this primal order for nature.

Polarities

If we follow the natural expansion of the vortex spiral, talked of earlier, we see the middle elements of Air and Water from level four, continue to divide forming a third polarity. We can identify the same divisions we found at the 4-fold level: While Dr Steiner describes the Macro polarity of the Outer planets as a Cosmic group, polarising to the Inner

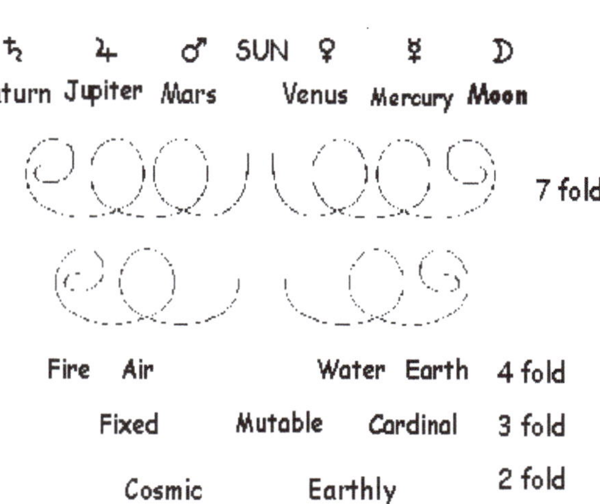

planets, Earthly group. We can also identify the Micro polarities between the individual planets of each group.

Macro-polarities

Because we are concerned with life on planet Earth, we observe the solar system as if we and the Earth are at its centre. From this position, as we look out we see there are three planets closer to the Sun than us, and three further away. These inner planets are the Moon, Mercury and Venus, while the outer planets are Mars, Jupiter and Saturn. The macro-polarity of the solar system is immediately obvious from this image. {See 3 top of pg 17}

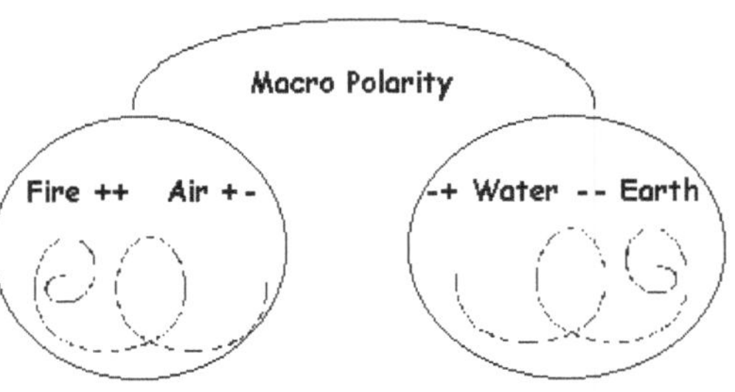

The outer planets are considered assertive and 'worldly'. Traditionally these three planets have been associated with the following attributes: Mars being actions taken in the world, Jupiter with the cultural and philosophic condition of any group, while Saturn governs the rules and bureaucracy of any social grouping. In Nature we see Saturn as carrying the Spirit impulse from the Stars, into manifest forms. Jupiter adapts this hard set of plans to the situation at hand, while Mars finds the appropriate physical avenues for these plans to be executed.

The inner planets are more personal and retiring in nature: Venus governs the receptive side of relating and what makes you feel good, Mercury generally covers communication, while the Moon covers the personal emotional response, instincts and nurturing needs. In Nature the Moon governs germination and the primary growth force, while Mercury plays the adaptive role of making the growth forces active in the plant through leaf formation, while Venus meets the outer planets from above and presents them with flowers to fertilise.

The outer planets are concerned with that which endures beyond the first cycle of any impulse, while the inner planets are associated with more transient phenomena. Each planet has its own individual quality, but also functions as an 'organ' in the body of the solar system. Indeed, medical astrology ascribes each planet the rulership of specific organs in the human body.

Micro-polarities

The micro-polarities found at the early layers of the Vortex continue as a development between an outer planet and an inner planet. These individual polarities form creative harmonies which can be identified in many different areas of life.

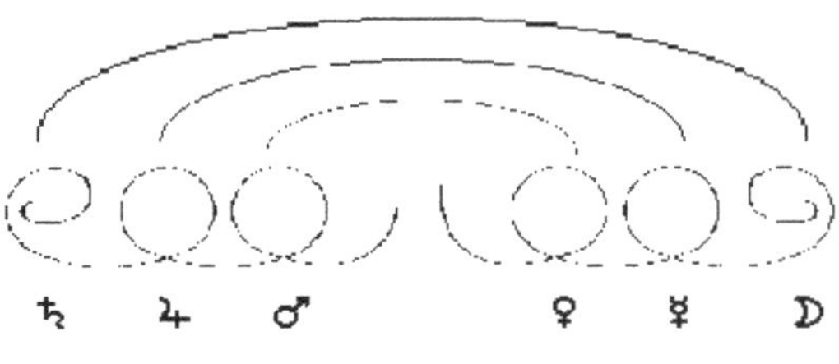

The Moon and Saturn.

The moon is the closest object to the earth, and is classed in Astrology as one of the planets. It is associated with the primary nurturing of and uncontrolled development of life. If left to itself, the moon forces would create life as an amorphous mass existing only to multiply as in primary cell division, hence creating the concept of 'The Blob'. It is emotional and 'watery' in quality and associated with a primary stage of intuition, through unconscious reaction to sensation. The first few years of a baby's life are very much moon years.

The Moon has the shortest cycle and is associated with quick growing plants, and those that retain a large content of water. On the one hand, this governs Cacti and succulents, and on the other it governs most of our vegetables.

Saturn is the wise old man, with an authoritative image, who brings structure and form to the Moon's growth. It brings form, manifesting as the human skeleton and skin, responsible action and social systems, or bureaucracies. Left to itself it would continue until we and its other forms, become sclerotic stalactites. It is only through the balanced interplay of the Moon and Saturn (the child and the adult) that life finds form and maturity, at their allotted optimum time.

Saturn rules the evergreen trees and especially the conifer varieties. Their rhythms are usually of long duration and their lifespans are in multiples of 30 years. The restrictive quality of Saturn is noted in their pyramidal shape. Their sap is slow and often resinous. This polarity is - the will to manifest.

Jupiter and Mercury

Traditionally, both Jupiter and Mercury are considered to govern intelligence. Mercury is quicker in nature, and scurries about collecting all the facts. Jupiter is the philosopher who examines, balances and gleans wisdom from Mercury's information.

The image of Jupiter is of the well-rounded mature entity, governing the cultural life of a community and the middle years of life from 30 till 50 years. This is the period when one has accumulated experience and wisdom and still has the health to act consciously. It shows in plants as well as humans. It governs the deciduous trees that often form well rounded spheres, for instance the English Oak, and fruit trees of all kinds.

Jupiter trees are deciduous and form spherical canopies, with their sap being more fluid and their lifespan usually of moderate duration. Fruit trees and the Oak family are good examples of Jupiter. The abundance of Jupiter is evident in the fruit we gain from her charges.

Mercury on the other hand governs climbing and creeping plants that need a supporting tree to hold them upright. The adolescent with their endless stream of unfocused energy personifies Mercury. The key thought here is that both planets expand whatever they touch.

Mercury runs and weaves following the path of least resistance. The plants it governs are the vines and the runners. Mercury's mark is seen in any climber that does not support itself.

These two can be seen as - the thinking planetary polarity.

Mars and Venus

This is the polarity of the feelings.

Venus epitomises all that is beautiful and harmonious in the environment. It can be said that its sole aim is pleasure, and being an inner planet, it gains pleasure through attraction i.e. by its beauty it attracts. It is concerned with an easy interchange between any two objects and aims to achieve harmony. It creates the milieu for social interaction to occur. Romance and foreplay are both Venus' activities.

Venus rules plants in which flowering predominates and has been associated with Alpine and desert plants. These plants lie dormant for long periods, and when the conditions are right, flowering occurs. Their flowering can be extremely short, but is the highlight of their life.

Mars, on the other hand, takes action to get what it wants, and is prepared to achieve its aims at any cost. With Mars there is always some loss in the process of gaining the prize. It is gross compared to Venus but in their intermingling a harmony is reached and Mars happily gets for Venus anything she wants.

Mars' cycle is 2 years and governs both biennial plants and the shrubs, especially the shrubs that have an element of die-back in their growth. After flowering and fruiting the new shoots come from the growth below.

In so doing these are the planets of relationship and feeling.

The Sun
The Sun is symbolically the central spiritual individual, the "I AM" presence, who mediates and harmonizes the creative tension set into play through the planets. As such it is a balance of all the above energies. The Sun is the mediator of all the planetary forces and its plants show a balance of all the parts. The clovers and grasses are Sun ruled. The roots, leaves and flowers all have equal predominance in the life cycle.

The "far-out" Planets
Uranus, Neptune and Pluto, being collective influences generally work in areas beyond the individual, in group phenomena. Uranus governs Group Ideologies, Neptune -Group Faiths while Pluto governs Group Movements. If an individual has personal planets closely related to these outer planets it gives them the opportunity to connect with collective impulses more closely. Telepathy and psychic perception arises from this situation. Uranus creates the Occultist, Neptune the Mystic and Pluto the Shaman.

They do indeed create 'far-out' individuals, some might even say " beyond the fringe".
These divisions give some hint as to why planting by the Moon is so popular among vegetable growers, as the Moon has the closest affinity with vegetables. Experience has also shown that emphasising the influence of other planets at the sowing, planting and harvesting of their associated plants, helps in their growth and vigour.

These associations are archetypal in nature and while the varieties mentioned, carry the mark of the planet most prominently, there are many plants, that carry a mixture of influences.

This planetary order is useful as an overall image of Biodynamics, however in the

SPIRIT Indirect Outer Pl. 1 Warmth Cosmic Forces Clay	**Cambium** **Salamanders** ↓	**Saturn**	P	**Valerian**	**Strengthens the Spirit**
	Sylphs	**Jupiter**	S	**Dandelion**	**Merge Spirit and Astral towards the Physical**
ASTRAL Direct Outer Pl. 2 Light Cosmic Substance Sand	**Life Sap**	**Mars**	Cl	**Nettle**	**Harmonise the Astrality**
		Sun	Ar	**Equisetum**	**Astral stimulates the Etheric within the metabolism**
ETHERIC Direct Inner Pl. 1 Chemical Earthly Forces Humus	**Undines** ↓	**Venus**	Na	**Yarrow**	**Opens the Etheric to the Astral**
	Gnomes **Wood Sap**	**Mercury**	Mg	**Chamomile**	**Stimulates the Etheric**
PHYSICAL Indirect Inner Pl. 2 Life Earthly Substance Cations	↑	**Moon**	Al	**Oak Bark**	**Etheric binds to the Physical**
	500	**Earth**	Si	**Quartz**	**Spirit binds to the Physical**

Agriculture Course we have to take another step to really understand what RS is offering, with his other planetary hints.

So if we keep things simple and disregard several annoying comments there is 'no problem', RS is simply describing how the planets relate to the physical bodies of the 4 kingdoms……until we look deeper and especially when we come to lecture 6, and diagram C.

In the 6th lecture, where we are dealing with **manifest** problems and the planets are reversed, the Outer planets are coming from above, and the Inner planets are coming from below. Along the way RS speaks of Direct and Indirect planets, in three different contexts. Given this confusion and small amount of text devoted to the planets, people easily dismiss these planetary indications as interesting curiosities, and choose one or other of the directions. Commonly lecture 2 wins, as this conforms to the basic Biodynamic image of plant growth, even though that order does not conform to the basic logic we found in Lecture 1.

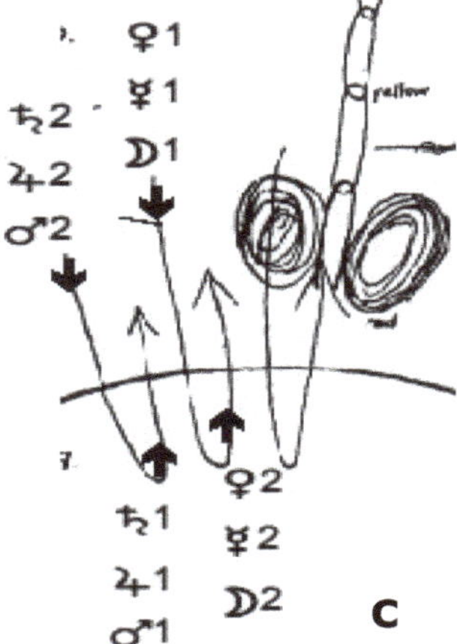

Being an Astrologer, all this means a lot to me, so losing the planets is not an option. Even to a reasonably astute and experienced Astrologer, Dr Steiner's use of the planets are challenging. He plays a few 'games' with them that are not commonly found in most Astrologers bags of tricks. He uses the planets more as part of a formula of forces, that can move into different arrangements according to their contexts, rather than as set of characters sitting around a table chatting – an imagination that helps in animating a birthchart. As unusual as this is, it is important we find our way through his games, as the planets are the golden thread through his story, that allows all the various layers and their parts, to be followed through this very winding road, we call Biodynamic understanding. I have been exploring these byways since 1976, with my Astrologers eye, so while my story may seem complex, it is coherent and the only way I have found, that all the pieces of the puzzle fit together, without the trickery of religious references.

3 Stages of Manifestation

Along the way I have created a map of sorts, that provides the contextual framework to what is best imagined as a 3-stage development of a 6 layered game of chess. Dr Steiner is outlining a process of manifestation, that does not take place all at once, and certainly can not be told in one sitting. Dr Steiner and Dr Lievegoed have clarified this 6 layered game clearly enough. However the 3 stages that this 'Creation's Vortex' moves through, is not so clearly stated. Thus few people have been able to follow 'the path' very deeply. While the 'dots' are throughout his literature, the pathway through those dots, is not stated. This pathway shows there are three stages for the process of the Universal Intention or Archetypal Thought to become Manifestation. This three stage process can be seen in the most fundamental of life processes, Cell Division. We begin with an existing sustainable cell (1), which then begins to pulsate for a time (2), before it then divides (3).

Similarly, I have come to see in many things there is an Archetypal Impulse, coming from the Stars; there is a stage 'behind manifestation'; and then there is manifestation. This may seem all very academic, however when we explore Dr Steiner's stories , we can find these three stages being outlined. If we do not know which stage is being discussed at which time, considerable confusion arises, and why 99% of readers believe the Agriculture can not be understood.

I believe it can be understood. The wrong tool has been used for the job at hand.

I first saw a hint of these three stages in my journey through 'Biodynamics Decoded' which I made available in the 1980s. I had realised that the perspectives used in astrology - the Zodiac , The Planets, The Elements, The Modes, Duality and Unity - were all different levels of the one Vortex. Each level describes the same thing - reality - but within the pattern and laws of its parts. Unity polarises to complementary opposites, male & female, positive & negative, etc. At the next level these polar aspects interact to form a middle space and together these are known in astrology as *the modes*: cardinal, fixed and mutable. In turn this middle divides into two to give a four-fold division, manifesting as the classical elements: *Fire, Air, Water* and *Earth*. From here the middle condenses and divides again and we have 6 pieces - the planets. At the wide mouth of the vortex the 6 planets each reveal two aspects, a positive and a negative side, to reveal the Zodiac.

All of this 'pattern forming' is reflected in our astronomical reality, and has been substantiated through 5000 years of astrological observations. This suggests we are observing a universal truth about reality. Rudolf Steiner's stories throughout his latter lectures, often talked of one or other of these layers of organisation. He provides the manifest flesh to these archetypal bones.

A question arose though. By following the progression through the vortex layers outlined above we come to an order of the planets Dr Steiner uses, based upon the length of their cycle, as we experience them. When we move to the Zodiac we order them according to their traditional planetary ruler and by their positive or negative grouping. The pattern of the Zodiac that arises from this process does not conform to the pattern of the Zodiac as the constellations appear in the sky. The vortex story told in 'Biodynamics Decoded' has the pattern revealed in the Step 1 picture. To get to the Step 3 picture, which is how the Zodiac is ordered in the sky, we first have to 'unwind' the mutable or changeable constellations, which provides an image of the constellations as a lemniscate, - which RS uses in his 12 Senses study — and then this lemniscate has to unwind another step to reveal the familiar zodiac, with its inherent complex of polarities.

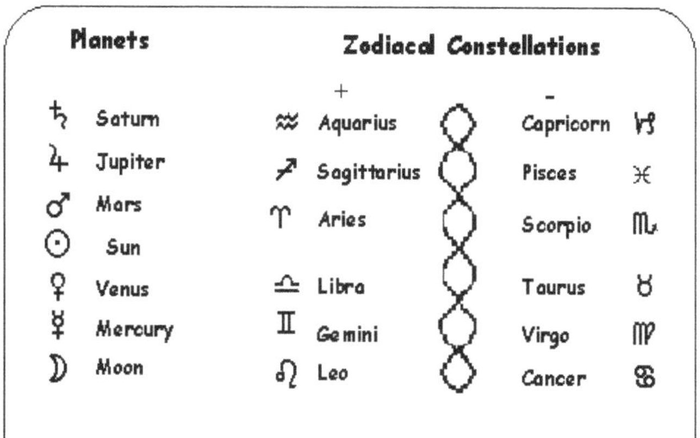

While we can see the 'normal' Aries > Pisces zodiac in this image, we can also see a zodiac that starts at Cancer - ruled by the Moon - and runs backwards through the planets till it reaches Leo, ruled by the Sun. This is the zodiac used with the precession of the equinoxes, that describes Human historical societal development. Dr Steiner and Eugen Kolisko also use this zodiac to describe

Planets		Zodiacal Constellations			
			+	−	
♄	Saturn	♒ Aquarius		Capricorn	♑
♃	Jupiter	♐ Sagittarius		Pisces	♓
♂	Mars	♈ Aries		Scorpio	♏
☉	Sun				
♀	Venus	♎ Libra		Taurus	♉
☿	Mercury	♊ Gemini		Virgo	♍
☽	Moon	♌ Leo		Cancer	♋

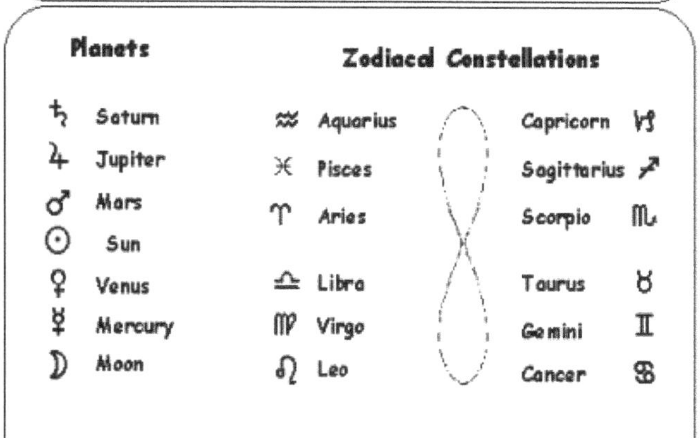

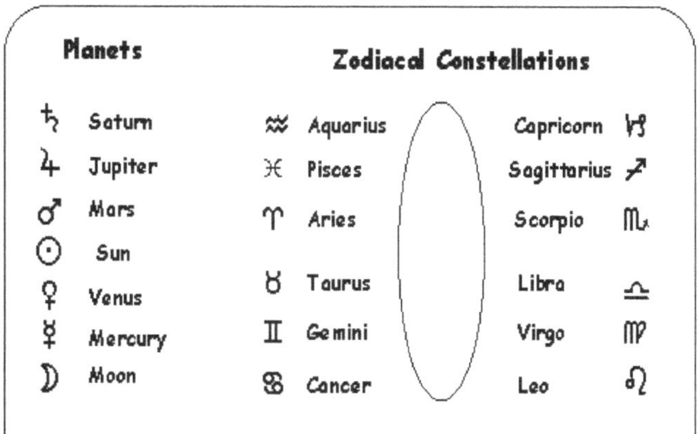

Step 1 — Zodiac

Planets		Zodiacal Constellations			
♄	Saturn	♒ Aquarius		Capricorn	♑
♃	Jupiter	♓ Pisces		Sagittarius	♐
♂	Mars	♈ Aries		Scorpio	♏
☉	Sun				
♀	Venus	♎ Libra		Taurus	♉
☿	Mercury	♍ Virgo		Gemini	♊
☽	Moon	♌ Leo		Cancer	♋

Step 2 — Zodiac

Planets		Zodiacal Constellations			
♄	Saturn	♒ Aquarius		Capricorn	♑
♃	Jupiter	♓ Pisces		Sagittarius	♐
♂	Mars	♈ Aries		Scorpio	♏
☉	Sun				
♀	Venus	♉ Taurus		Libra	♎
☿	Mercury	♊ Gemini		Virgo	♍
☽	Moon	♋ Cancer		Leo	♌

Step 3 — Zodiac

the archetypal order of the evolution of the animal kingdom, and I have found this zodiac useful in understanding chemistry.

When I came across this pattern and the threefold unwinding process, I supposed that it was important, but I did not know why or how. So I 'noted it' and put it on the shelf. Subsequently, I came across similar things through the years.

Once I thought I saw RS describe this whole process, but I have been unable to find it again. I have asked various students of Dr Steiner if they have seen this pattern described but none have. I suspect I must have seen this in a dream or the like.

The next place I came across this 3-stage process within phenomena was when I began to study Chemistry. (1)

Stage 1

One of the first observations one can make about the chemical elements is that they are organised into 7 layers as revealed in the periodic table. When we look at the organisation of our creation, we can note that we have 6 layers of immediate organisation. The Earth, the duality of matter, the physical organisms that have a 3-fold organisation, into the nerve sense, rhythmic and metabolic system of a physical body, the atmosphere of the Earth—where the 4 elements exist, the 7

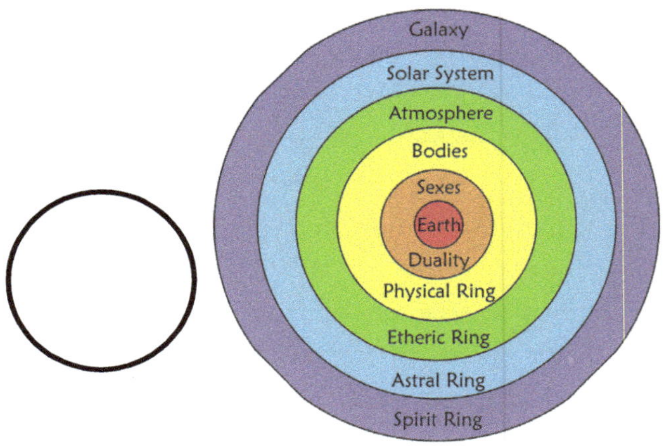

planets we can see with our eyes, and then the galaxy of stars. This is our immediate environment. The 7th layer is either the rest of the Universe, or we can look at the Zodiac layer as having two layers within it. I choose this latter option. In the layers of electron shells of chemistry, we have an organisation that is a reflection of what is real in our environment. As above, so below.

This organisation provides a clear image of the Cosmic activities. Dr Steiner states in various places, that these cosmic spheres are the actual source of the activities we call Spirit, Astral, Etheric and Physical. The stars are the spirit forces, the planetary spheres are the astral activities, while the atmosphere of the earth is the world etheric, while the Earth provides us with the physical substance that these activities use to become manifest as organisms.

Stage 2

The universe is not stationary however. It is moving and spinning very fast. When something moves there is polarisation of its matter into positive and negatively charged elements. This is due to the formation of a magnetic field and its subsequent electrical field. This polarisation leads to life and to the specific ways the world bodies outside us and internal bodies inside us organise within life. So far in the astronomical story we have four players, but once things being to move two more layers are created - polarity and the 3-fold physical bodies that manifest from this polarity. This physical body exists within the four external spheres.

In chemistry, polarity shows as the cationic and anionic 'families of elements', of which there are 8 families of the major elements. Through studies of the nature of the families, much helped by Dr Hauschka's book, (2) I was able to organise the major families into the 8 arms of the octagonal gyroscope. This is a form used by most cultures of humanity in the building of their temples and cathedrals. The final clarification of this double cross image was my consideration, based upon Rudolf Steiner's 1922 medical lectures. (3) This image as a structure for the chemical elements, provides a lot of insight into how they interact with each other and their correlations with the energetic activities.

Dr Hauschka provides a 'proof' of **the lemniscate as a feature of Life activity** in his chapter on the 'Brothers of Iron', where he tells the story of the transition elements in chemistry. The transition elements find their usefulness as catalysts of the biochemical life processes. While the major elements provide the raw material of lifeforms, the transition elements are essential for the inner processes of life to function.

Dr Hauschka points out there is a very odd feature in Mendeleev's classical periodic table of elements: The hardness and melting points of the elements increase with increasing atomic weight until the transition elements in the 4th layer of the table. Instead of the hardness and melting increasing with the order of the elements calcium, scandium … and so on through to zinc, these characteristics jump from calcium to zinc at the end of the transition group, then work backwards through to Scandium, and then revert to type with Gallium (see pg 43 of 1).

This dynamic suggests a better image for the Periodic table — which should be drawn as a circle and sphere rather than a rectangle given that atoms are spherical — shows a lemniscatory 'detour' within it incorporating the life-enabling transition elements.

Stage 3

While this arrangement offers insights into how all of chemistry interacts, there came a question (from Dave Robison) concerning its compatibility with RS's assertions in lecture 3 of the Agriculture Course that H, N, O, C are the carriers of the energetic activities into manifestation. These are the 4 base elements for all carbohydrates and proteins. Again note we are dealing with **real-life substance** — protein.

The answer to this question came by looking at where the elements are placed on the circular periodic table. Each element sits at the bottom of one arm, whose positive and negative companions form one axis. So this provides four axes of a positive and negative dynamic, within manifestation. This suggests a third layer of organisation by which the Periodic Table could be observed.

A further observation along this line came when considering the very physical realities of magnetism, electricity, electro-magnetism and di-electrics. Following the physical laws of their relationships, as described in their literature, these appear to follow the same pattern, as for the 4 base chemical elements.

Both of these references are of real manifest phenomena

So again I had a threefold process of the 'What Is', moving through a middle stage, into manifestation.

I then found a second Stage 3 'doorway' into Alchemical Chemistry (5). So I went hummmm, there is a pattern here. The Seasons investigations then showed me a third application of this organisaton.

Given this 3 stage process is found in the very basis of life's processes, cell division, we can consider this as one of the universal models, we might be able to trust.

The Double Planets

Rudolf Steiner's planetary organisation as described in the Agriculture Course is a very sophisticated thing, astrologically. While most Astrologers will recognise the 7 planets pattern, not many astrologers consciously work with incarnating and excarnating planetary processes, neither do they consider polaric groups working together as direct and indirect complexes, nor do they use the Cancer-to-Leo zodiac. Indeed, very few astrologers consider 'nature' at all. It is a very person-centered science in our times. Nevertheless, basic astrological experience and an awareness of older astrological traditions are very useful tools for understanding where RS is going.

I have already said Rudolf Steiner only left a 'breadcrumb trail' for us to follow,

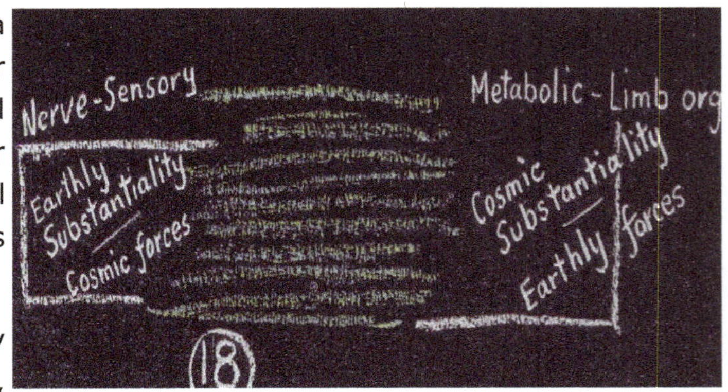

and the best diagram to begin this treasure hunt was provided in the last lecture of the Agriculture Course (pic 18).

The story of this picture was first given in Lecture 2, where we are introduced to the 4 Physical Formative Forces (PFF), working within the 3-fold physical bodies. RS only gives half the planets' story here though. He said the 'Cosmic' outer planets come from below, and the 'Earthly' inner planets come from above. The other half of the planet's story, where the outer planets come from above and the inner planets come from below, is given in the 6th lecture.

This lecture 6 order appears similar to what we can read from Lecture 1. However, Lecture 1 is the story of the outer World processes. In the lecture 1 context, the inner planets do work with the Earth and the Outer planets work with the Light and warmth in the Atmosphere. It is appropriate in this context to use the single line of the 7 planets as their cycles describe them.

Lecture 2 is the story of the internal forces active in 'living nature', and especially how the plant resides within and above the soil. After the first page, where we are told of the farm individuality, we are led into the 3-fold image of the plant within the Earths Physical body. This is where we are given half the planetary references, before Dr Steiner gives us the 4-fold story of the Physical activities working through the seasons. He begins with how the cosmic activities are being draw into the Earth via the siliceous sand, as Cosmic Substance. During the winter the Cosmic Force side of this activity is **Fructified** with the female Earthly Substance, which is what has come to the soil from the previously growing plant, and other local Earth forces. The task of joining these two streams is facilitated by the Gnomes (6). It then is sent upwards by clay, as entwined Cosmic Forces with Earthly Substance, (called Wood Sap). It is on the ninth page of the lecture that he tells us of the Earthly Substance and Force story, along with the physical substances that control

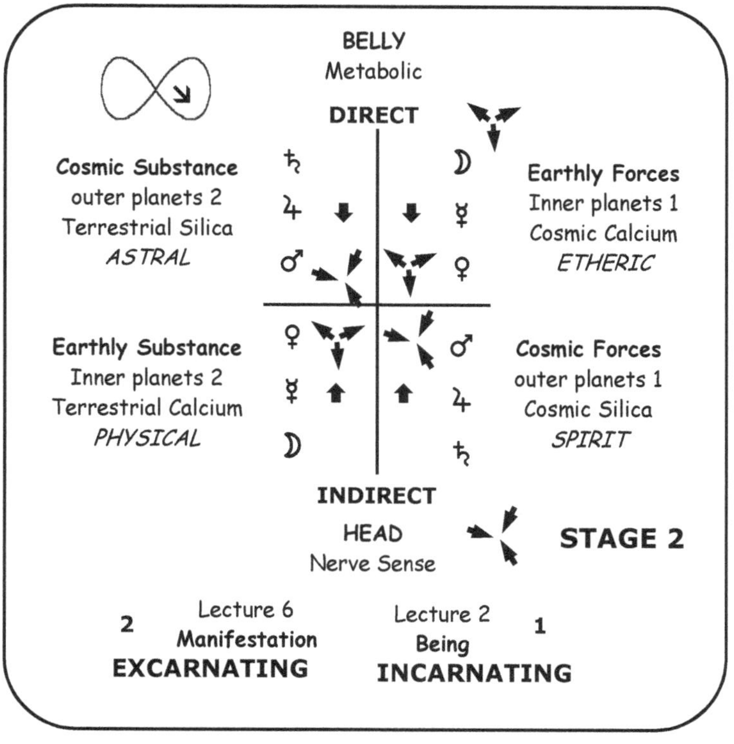

these processes actions. Firstly we hear how Lime and the Earthly Substance draws the Etheric into the Earth and then how Humus holds and strengthens the Earthly Forces working into the soil and plant. We then have a few pages telling us how these four activities all work together. - **clay, sand, lime and humus, all working as 'the middle' between Calcium and Silica.** These are activities working within the physical body, and can be viewed as similar to the 'organisations' RS talks of in his human medical lectures. (24) We can thus talk of a plant's warmth organism, being associated with the workings of the Cosmic Forces, and a manifestation of the Spirit working into the physical body, just as we can do with humans.

Lecture 6 tells us of manifest problems such as pest and disease, and there Dr Steiner outlines the 'substance' side of the planetary story to highlight this.

If we combine the images from lectures 2, 6 and 8 we get the picture at the bottom right of the previous page. We can also note this is the Stage 2 diagram. There is a lot of polarisation here.

The trick is to use both lecture 2 and lecture 6 indications of the planets, as parts of two circling interacting streams of activity. The Cosmic Siliceous circling anti clockwise, and the Earthly Calcareous circling clockwise. We can note this ordering is the same Dr Steiner uses for understanding the energetic make up of Humans.

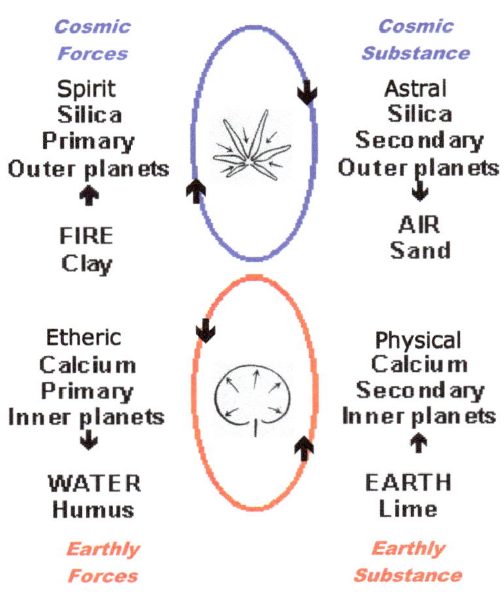

This is the answer to our central problem, but how we get here and how we can understand the comments about direct and indirect activities has its own story. So let's start at the beginning, as a preliminary step is required if we are to 'keep things tidy'.

Stage 1 - Lievegoed

RS's first premise is 'As Above, So Below'. ….. Life on earth is an expression of the Forces coming to Earth from the Cosmos, which are then reflected back outwards, as Lifeforms. The planets in this order are as they sit as rulers of the zodiac constellations. see page 30.

In 1951 Dr Lievegoed gave a series of lectures to the BD experimental circle (the Euro in crowd), which has been

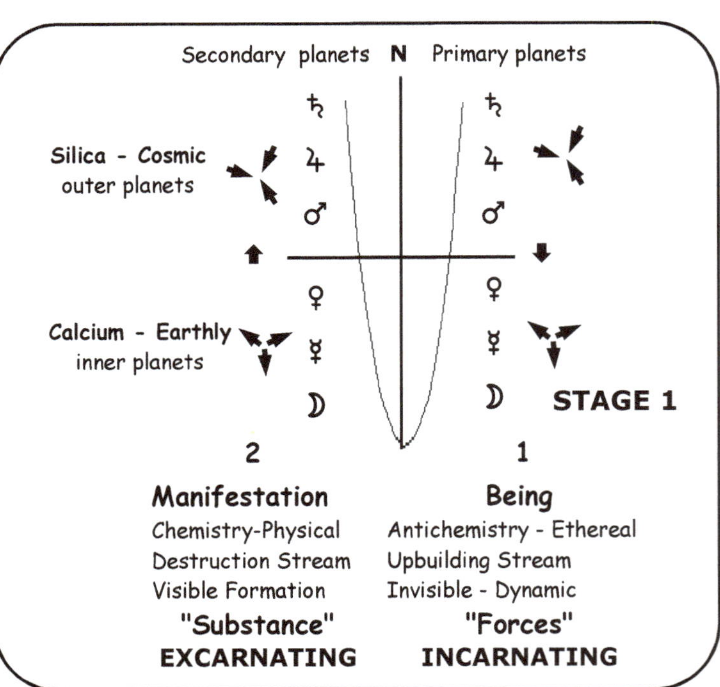

available as a booklet (4) ever since. It is not an easy book to understand, however it does offer further insights into RS use of the planets activities. This book is a gift from the Medical section to the Biodynamic section. This is how they understand the planets.

The planetary organisation Lievegoed uses is the 7 fold one we have discovered, but it is in movement. It is the 7 planet order coming inwards to the Earth and moving outwards from the Earth back to the Cosmos. It is the stage in between the simple order used in the course, ala Lecture 1, and the ordering used from lecture 2 onwards in the Agriculture Course. This is the same pattern RS uses for his 'evolutionary story', of how we take Birth, move through Life, then through Death and into Life again. This story tells of how, after Death we move through the planetary spheres, and undergo a process of reflecting upon our Lifetime. We do this, until we reach a point, just after Saturn, where we have a chance to decide if there is anything we would like to change. Upon this decision we then begin the incarnating journey through the planetary spheres, where we gather together the forces and activities we will need in our next life. We are born and then go through Life as a process of excarnation, living through the circumstances we have setup for ourselves. We are living out our life path, and towards our next death.

Plants have a similar journey. A species plan comes from some particular star, to the sphere of Saturn. Saturn picks up this 'idea' and gives it to the plant at Pollination. This plan is worked upon with Jupiter to adapt it to the circumstance and so on through all the planets, until the end of the incarnating Moon sphere stage., Here at Moon 2 germination takes place. Plant growth to seeding is then the excarnating phase of the run through the planets.

At this stage of the story we can talk of **Incarnating (1) and Excarnating (2)** planetary processes, however the Direct and Indirect processes are yet to come. See page 34.

In the second section of his book BL talks of planetary polarities, between these two streams and as opposites across his diagram . It is here where Lievegoed meets the Agriculture Course., however BL does not draw the pictures of these relationships in the same way RS does. BLs polarities are presented as oppositions, across the circle from each other. RS places them next to each other, and within the physical system that they predominately work. BL does not relate his planetary polarities to the physical systems of the Metabolic, Rhythmic and Nerve Sense as RS does in Agriculture.

Stage 2 - The Agriculture Course
This is the stage where we look at the influence of movement, upon what is there. With Movement things polarise. In the second part of BL's book, he talks of how in living processes, these planetary activities polarise with each other, to form a unified outcomes eg, Saturn 1 and the Moon 2 work together to create the skeleton and skin, and so on. (see pg 37) This is where the Agriculture Course takes over. RS gives us information about how the Soil, Plants and Animals organise and manifest according to the same planetary polarities, however he describes them in a broader context. BL's polarities become RS 'pairs'. RS places the planetary activity within the physical systems. In lecture 2 we are told how the root of the plant are like the head of the

Human, with the outer planets being active. While the leaf region is like the Human chest region, and flowering and fruiting parts of the plant are like the Human metabolic region, with the Inner planets being active. Within each of these physical zones we have to find each of the primary four activities of Spirit, Astral, Etheric and Physical working through their representatives at each layer.

RS story has 6 layers, but all the activities he mentioned are based on the four primary activities, running through all the layers. As the planetary activities work down into matter he gives them particular names. Once they work within the Physical organism, he calls the activity within the Metabolic system a combination of the Outers 2 Cosmic Substance and the Inners 1 -Earthly Forces. While in the Nerve Sense system or inside the Earth it is the activity of the Outers 1 -Cosmic Forces working with the Inner 1 - Earthly Substance as seen in the picture from lecture 8 , on page 23.

All these pieces have to be placed within the larger context, that includes the other inner layers of the story, as well as the external environment, to give us this image.

It is at this stage of our journey that BL and RS begin to talk about **Direct and Indirect Planets.** These are a Stage 2 phenomena.

The main difficulty of this topic, is that RS described three different versions of what constitutes Direct and Indirect planetary activities. I will go into the details of these three approaches in the next chapter.

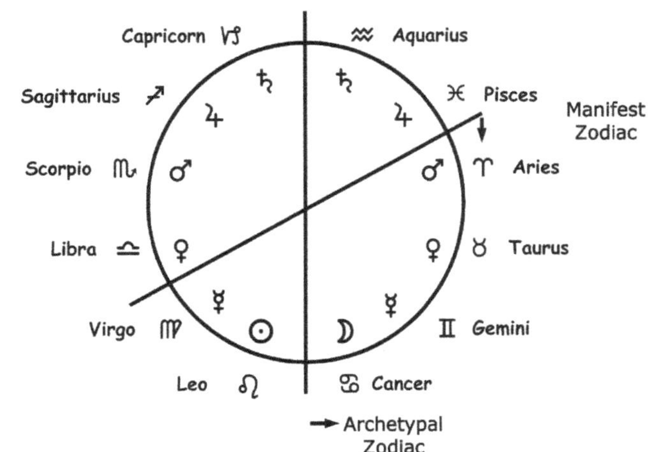

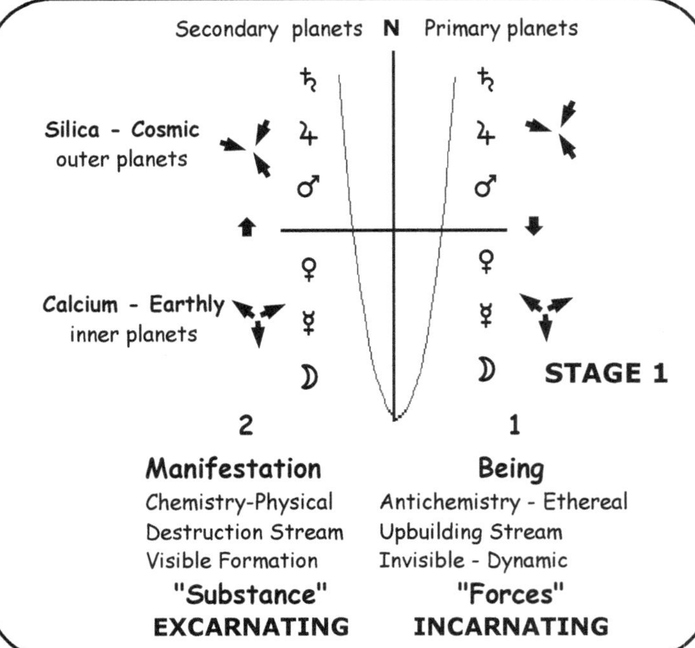

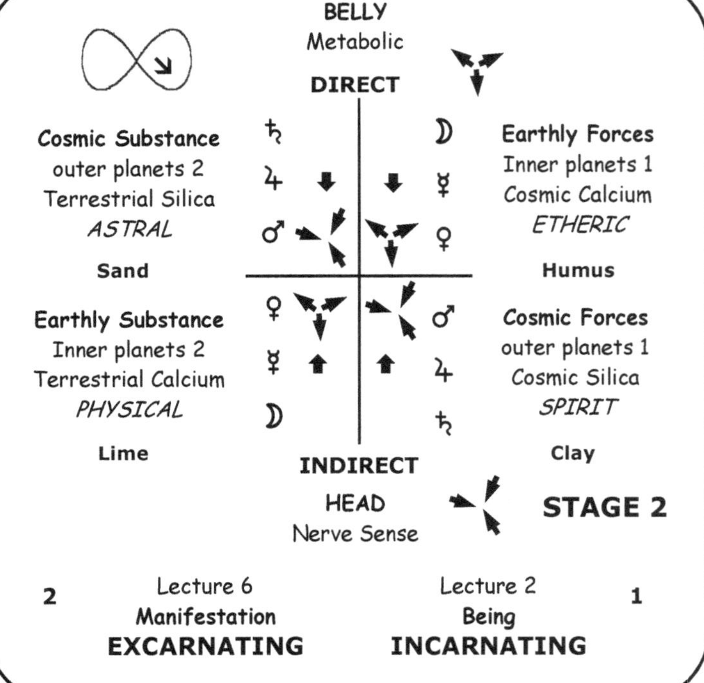

Each of these three organisations, appears to have their specific place of application. The one we are discussing here — where the Star and Planetary forces enter into the Earth as Direct Forces and are then reflected back outwards, as Indirect Forces, within the same basic geographical region — appears to be most related to the internal growth processes of living beings. These are firstly 'World' processes, occurring outside in nature, and secondly they work inside lifeforms, to influence the way internal activities occur.

With reference to this picture. The general Cosmic Silica and Earthly Calcium streams, are the circles in the middle. I see them as two streams of activity that spin in opposite directions to each other, moving between the Above and Below of the Earths surface. They are always interacting with each other.

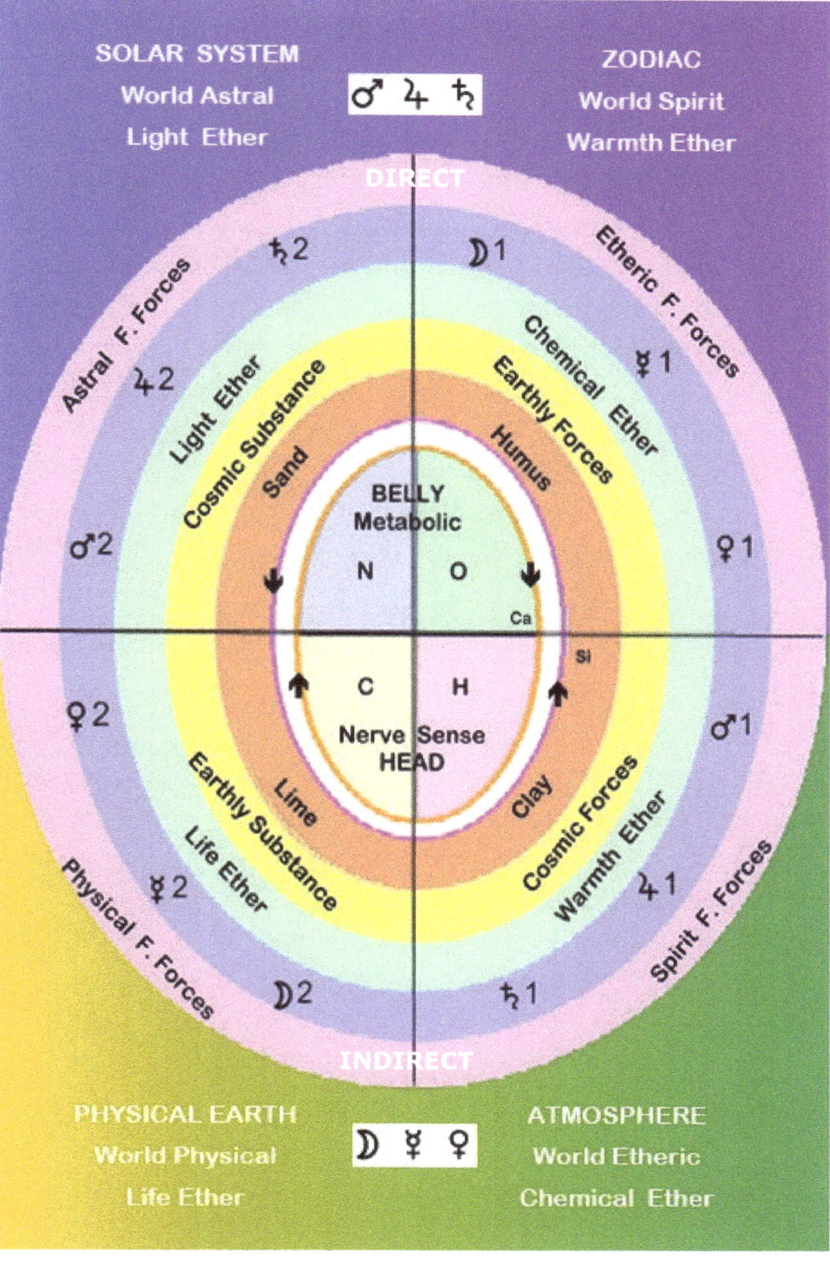

The individual energetic players of each part takes dominance depending on where the stream physically is. Eg **Below**, the Cosmic stream is dominated by the Spirit, and the Astral plays a secondary role , while **Above** , the Cosmic stream is dominated by the Astral and the Spirit is along for the ride. These two sides of the Cosmic Silica stream, have to interact with each other. The same occurs with the Earthly Calcium stream. When in the Earthly Substance quadrant, the Physical is dominant and the Etheric supports it, whereas when the Calcium process is above the soil in the Earthly Forces phase, the Etheric is the dominant activity with the Physical playing a secondary role. This is the Biodynamic Plant Dance. **How do we work with all these processes to enhance Nature. We can use the 4 Earthly substances, clay, sand, humus and cations, along with the BD preps.**

We can look at the Layers of the Quadrants, as different dimensional manifestations of the one energy. They are different shells within our the Internal Universe, that can each effect the whole quadrant. Strengthening any one part of this activity, and the whole quadrants activity is influenced. Biodynamics offers us the opportunity to know the parts of the game, and what will influence them. We are particularly interested in the Planetary ring in this discussion. If we use the Outer planets 1 quadrant as the example. The two main players driving this area, are the Spirit and his mate the Astral. When they are active within the planetary dimension, Saturn expresses the Spirit, while Mars expresses the Astral and Jupiter is how these two activities work together. Being of the primary phase, Saturn holds a Star's Species message for the plant. Jupiter adapts this primary impulse into something that can fit into the environment it finds itself in, while Mars gathers the enthusiasm for the task, assesses whether these goals are possible, based on Jupiter's primary research, and begins to move forward. This is the force that ,once it has grounded itself through its time in the Earth, wants to bolt to seed, in the spring, to further its kind. All the other forces have to hold on for the ride and try to soften its will enough, to build some leaves and fruit.

To get from Stage 1 to Stage 2 there is a twist needed in the diagram. On the Incarnating side, we flip the planets over. This provides RS 'pairs' within the physical systems. Once we enter into Life Processes everything is polarised. Cations align with Anions, Positive and Negative charges attract and so on. So , like the pulsating / polarising phase of cell division, this movement towards Life appears in the diagrams via a Lemniscate twist. This middle stage 2 stage provides, the laws that stand BEHIND creation.

The Incarnating and Excarnating Streams, are still in place, albeit lemniscated. The Direct and Indirect activities are only talked of in regards to the Stage 2 and 3 organisations. (See BL pg 19). While Stage 2 is a development of Stage 1 these 'Stages' work simultaneously, and independently of each other. Two distinct dimensions with different 'rules of the game', in the same space. I imagine these 3 stages being related to the three planes of the Gyroscope. Stage 1 is the Vertical Magnetic plane, Stage 2 is the Electric Horizontal plane and Stage 3 is the Electro-magnetic / DiElectric, vertical plane. 3 planes of activity in the same

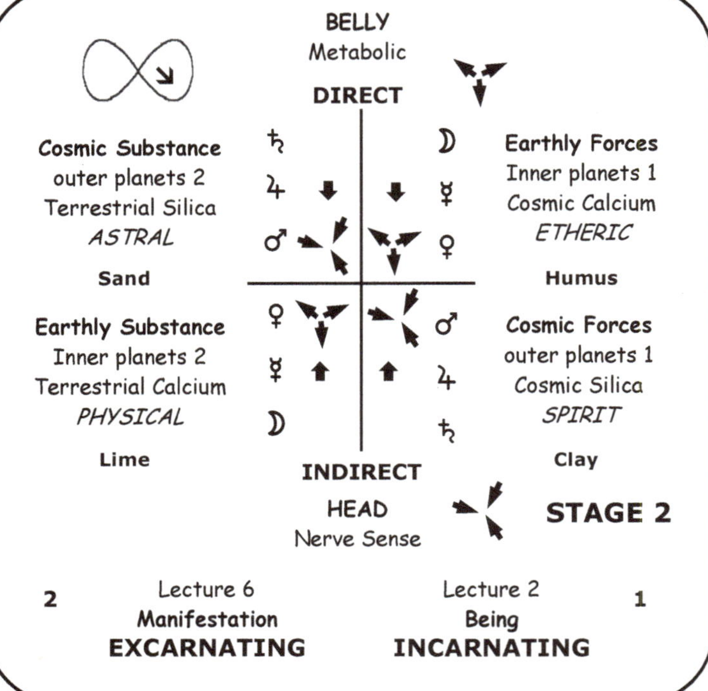

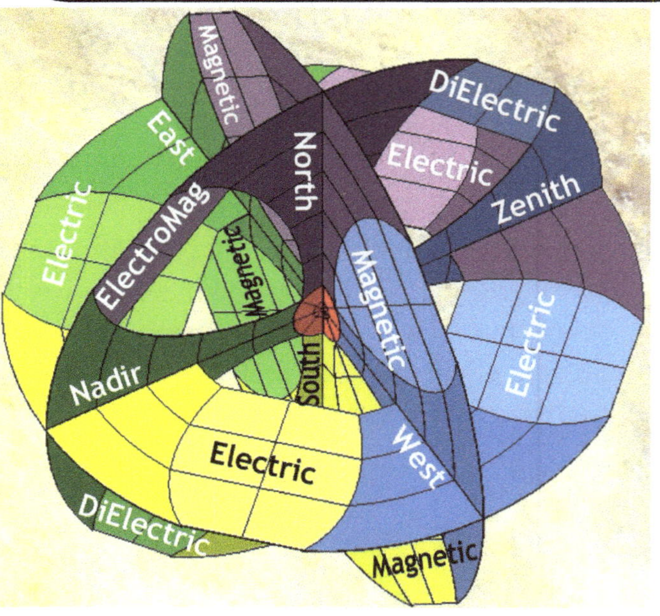

place. This says Stage 2 - the Agriculture Course is the story of the horizontal plane, upon which we find ourselves, along with the planets in the Solar system, and behind which we have the constellations. This is the plane and place where life exists. These are the processes that bring Life into Creation.

When we put lectures 2 and 6 together, we can see RS has the **Direct** Forces coming from above, and working into the Metabolic processes, which has two sides. These are forces coming from the Outer planets 2, (those beyond the Sun) which have a inward moving contractive Catabolic action called Cosmic Substance ,and the Inner planets 1, (those between the Sun and the Earth) referred to as Earthly Force activities, with their expansive Anabolic quality.

 The **Indirect** Planets are coming back from the Earth, and are made up of the Inner 2 called Earthly Substance and Outer Planets 1 called the Cosmic Forces, working from Below. Thus we have a expansive and a contractive influence in both the Inward Direct and the Outward Indirect activities. We can also say there is one Incarnating and one Excarnating process in the Direct and Indirect streams of activity.

So while we can find a dominant Etheric expansive process in the Metabolism, we can also find a contractive Astral growth process, working with it. Similarly within the Nerve Sense system, it has a dominant contractive influence, supporting the contractive Spirit, based Cosmic forces, with a secondary expansive influence, coming from the Physical Earthly Substance processes. This expansive influence has 'migrated' from the metabolism, into the head, through evolution, and still works upwards to support the 'nourishment' of the brain substance. Too much of this and we have migraines.

Finding the details of this organisation occurred over several years, and came from a few lines of observation coming together. No doubt if I had some external education from a Apop doctor I would have come to it much quicker, but this has not been my path. Nevertheless here it is and this is what you will find Apop medicine uses as a basic reference frame of activity.

From stage 2 we have the **Direct Physical Formative Forces (PFF) Cosmic Substance (outers 2) and Earthly Forces (inners 1) in the Metabolism, and the Indirect Cosmic Forces (outers 1) and Earthly Substance (inners 2) in the Nerve Sense system**.

I was not firm on each of these groups 'needing' to sit in any particular order, in my early diagrams of this, however as things started to coalesce, it suggested where each of these sit in RS lecture 8 fourfold pattern, is indeed significant. Dr Steiner's diagram placements are to be taken seriously, for Stage 2 where the forces active behind manifestation are described. There is however a further step to take into stage 3 , Enfolded Manifestation. This is not found in the Agriculture Course, but can be found described in 2 other lectures (6) as well as in Chemistry and the Seasons.

Before we go there lets finish off the Direct and Indirect stories.

Direct and Indirect Processes

Apart from the previous story of how I see RS talking about the Direct and Indirect activities, there are two more ways RS tells us of Direct and Indirect planetary activities are possible.

As part of the basic axiom 'As Above , So Below' RS starts with the polarity of the Cosmos beaming forces from Above, and the Earth receiving these, and then reflecting them back outwards. A simplistic Cosmic and Earthly designation does not allow for the required complexity we have later in the story, so he clarifies forces from Above as being Direct, and forces coming from the Earth as Indirect. But exactly when is this clarification applied?

With the Indirect activities there are three variants that he mentions. (A) is in Lecture 4 where the Direct Sun or other planetary forces are reflected off the Moon or other planets before reaching the Earth as Indirect forces. Thus we have the Direct Sun and planetary forces and then we have Indirect Sun and planetary forces via the Moon and other planets.

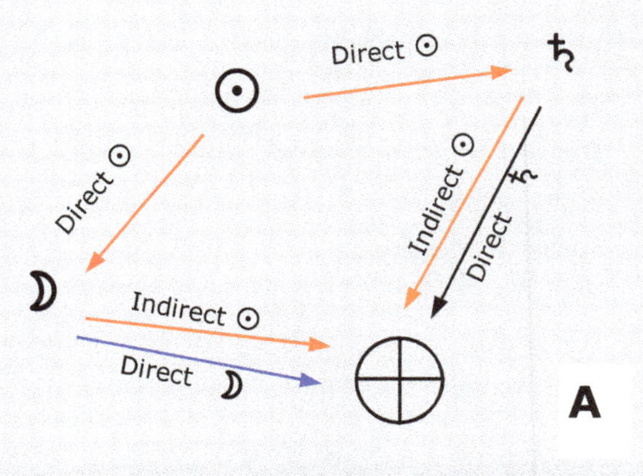

(B) Is where the planetary influence comes into the Earth from the opposite side of the planet to the plant or animal. So the Direct comes from Above and the Indirect comes through the Earth. We need to note that this will occur each day, as the Earth rotates. This is talked of in Lecture 1

Both A and B forms of Direct and Indirect are talked of in relation to animal formation.

(C) is when the planetary influences come directly from above, and are accumulated in the Earth, before being released back upwards. While this process is a continual process, it also has a seasonal component to it. As part of RS's 'Seasonal' story, some of which is told in the second lecture, we are told the Direct planetary influences, especially the Cosmic Outer Planet Silica processes are drawn in through the Autumn, 'crystallised' at mid winter, and then released back upwards, as Indirect forces in the Springtime. This is talked of in lectures 2 and 6, in regards to plant growth. They are also talked of elsewhere.

(b1) Astronomy Course -lecture 7 -7th January 1921. http://wn.rsarchive.org/

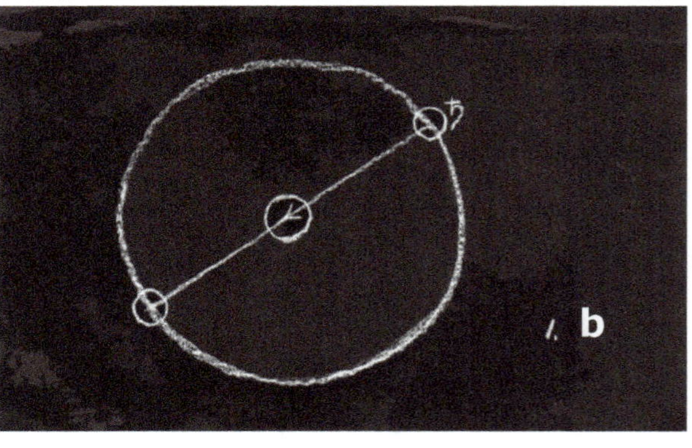

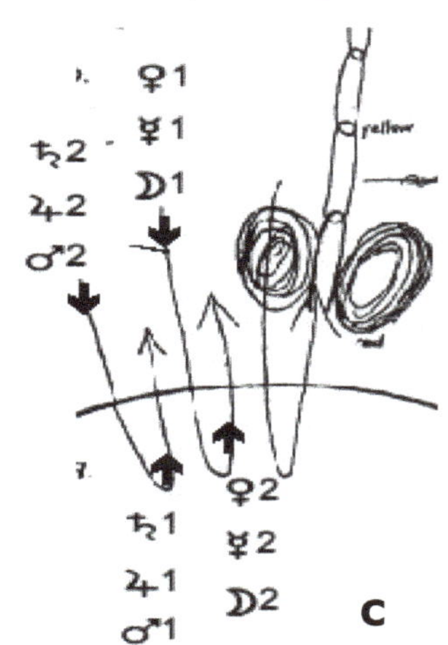

Lectures/Dates/19210107p01. html
(b 2)Human Questions Cosmic Answers lecture 4 2 July 1922 https://wn.rsarchive.org/
Lectures/Dates/19220702p01.html
(b3) Agriculture Course lecture 1 — 7 June 1924 http://wn.rsarchive.org/Lectures/
Dates/19240607p02.html
(c 1) Agriculture Course lecture 2— 10 June 1924 http://rimu.geek.nz/garuda/
Agriculture/Lec%202%20finish.pdf
(c 2) Agriculture Course lecture 6 — 14 June 1924 http://wn.rsarchive.org/Lectures/
Dates/19240614p02.html

Summary of texts

(b1) "Diagrammatically now let this be the animal form. If after going into an untold
number of intervening links in the investigation, you put the question: 'What is the
characteristic difference of the front and the back, the head and the tail end due to?',
you will reach a very interesting conclusion. Namely you will connect the differentiation
of the front end with the influences of the Sun. Here is the Earth (Fig. 3). You have an
animal on the side of the Earth exposed to the Sun. Now take the side of the Earth that
is turned away from the Sun. In one way or another it will come about that the animal
is on this other side. Here too the Sun's rays will be influencing the animal, but the
earth is now between. In the one case the rays of the Sun are working on the animal
directly; in the other case indirectly, inasmuch as the Earth is between and the Sun's
rays first have to pass through the Earth (Fig. 3).

Fig. 3

Expose the animal form to the direct influence of the Sun and you get the head. Expose
the animal to those rays of the Sun which have first gone through the Earth and you
get the opposite pole to the head. Study the skull, so as to recognize in it the direct
outcome of the influences of the Sun. Study the forms, the whole morphology of the
opposite pole, so as to recognize the working of the Sun's rays before which the Earth
is interposed — the indirect rays of the Sun. Thus the morphology of the animal itself
draws our attention to a certain interrelation between Earth and Sun. For a true
knowledge of the mutual relations of Earth and Sun we must create the requisite
conditions, not by the mere visual appearance (even though the eye be armed with
telescopes), but by perceiving also how the animal is formed — how the whole animal
form comes into being."

(b2) Now super-sensible observation has revealed that everything that radiates from
the light of the moon, everything that streams as reflected sunlight from the moon on
to the earth, and also streams into our thought-life as formative force — all this works,
too, in the shaping of the animal forms. Essentially, all that is indeterminate, formless
will-force in the animal is to be found within the sphere of the direct light from the sun.
But all that gives the animal its independent form, which is not adapted to the earthly
element, is, in the true sense of the word, woven out of the gleaming moonlight."

We need to note here, the animal Head is influenced by the Direct forces, while the Metabolism is effected by the Indirect forces.— Either Sun forces through the Earth or reflected by the Moon. When this relates to the plant though as in scenario C, it would seem it is the other way round. The Roots / Head will be influenced by the Indirect forces, and the top of the plant by Direct forces.

(b) "Saturn goes slowly round, in thirty years. Let us draw it thus (Diagram b): here is the course of Saturn. Sometimes it shines directly on to a given spot of the Earth. But it can also work through the Earth upon this portion of the Earth's surface. In either case the intensity with which the Saturn-forces are able to approach the plant life of the Earth is dependent on the warmth-conditions of the air. When the air is cold, they cannot approach; when the air is warm, they can." Hugh Lovel refers to this in some of his writings, (and from memory) he does not differentiate between this and C.

Enzo Nastati also discusses this subject, however I do not agree with his view of Direct and Indirect. He believes the Incarnating and Direct activities are the same, while the Excarnating and Indirect are the same. He does not talk of them as I do.

The Double Planets and their Constellations

Dr B. Lievegoed, D. Robison, Dr G Atkinson

In this chapter, David Robison's adaptation of Dr Lievegoed's text
is in normal font and my additions in italics

As an adjunct to all that has been said about the double Cosmic and Earthly activities, we can deepen this discussion by an investigation of the double planetary activities, which were only touched upon in the Agriculture lectures. While BL gives a chapter on the planetary relationships of the BD preparations, he does not clarify how the preparations work with the double planetary activity, or how we might emphasis one process or the other. Given each planet rules two constellations, it would seem reasonable to explore the effects of using the preparations in one or other of its constellations, to see if this might focus upon the Incarnating or the Excarnating planetary process. First step though is to clarify which constellation fits with which planetary process. I do this with the help of Dr Hauschkas correspondances of the chemistry to the constellations and my own Astrological experience. (2)

Organisms manifest *physically* in a threefold way; for the plant, the root, leaf, and flower correspond to the head, chest and limbs in animals. *The* Etheric *processes* demonstrate four types, corresponding to the four states of matter, *Warmth, Light, Moisture and Substance.* An organism does not come into being merely through etheric processes, but only where *the spiritual archetype directs the* astral principal *to form the etheric activities, that mould the physical.* The *spiritual* archetype induces the threefoldness based upon an opposing polarity, with a middle rhythmically linking the poles. Non-living things may have polarity, such as a magnet, but they lack the threefold organization, *we had described earlier.*

The astral principle is carried as a moving, dynamic activity based on a sevenfold archetypal pattern. These seven working principles influence the formation of specific organs in the animal, from which their influence permeates the whole creature. Spiritual principles are archetypically arranged by twelve principles *that are sourced from* the zodiac. These principles are responsible for the specific individuality (Ego), that which makes a plant be a rose and not, say, sage. The seed is the carrier, or point of attachment, for the spiritual forces that define the species.

The seed only unfolds when surrounded by the physical, etheric and astral elements in such a way that these forces can all stream freely into one another. This is what fertile soil provides. Think of a super-saturated salt solution; just a small crystal is enough to induce the whole solution to crystallize. Similarly, a truly fertile soil "longs" to be a plant and is just waiting for the contact of a seed to give it an archetypical "instruction set". It's our job to make the "pre-plant" condition.

We do that with compost *and remineralisation.* Compost is formed from decaying plant and animal substance, so it still has something of their physical, etheric and astral forces. But these forces are dispersing and the spiritual forces have already withdrawn.

We try to capture the dispersing forces and bring them into cooperation once again so that new spiritual archetypes can be received. The remnants of the old organs (leaves, etc) are to be destroyed but their etheric and astral principles are to remain united in the compost. Imagine a caterpillar in its cocoon; it must go through a period of chaos before the new body of a butterfly can form. Likewise the compost goes through chaotisation, in order to be at the disposal of the seed. But there is an essential difference - in the animal, astral forces work from within, concentrated in the organs. While in the plant, astral works from outside, streaming through the plant. To conduct the chaotising harmoniously, we implant something like planetary organs in the compost pile.

We do this by inserting the biodynamic preparations. Six of the preparations represent specific planetary forces. The seventh is the balancing force of the sun, represented by the horn manure/horn clay preparations. The farmer should be conscious of these underlying processes and not just operating   mechanically or dogmatically. The place to learn about such physiological processes is from the doctor. In fact, the farmer can learn much by considering how the physician treats the human body. And the doctor can enrich his knowledge from the observations of biodynamics.

The reader should be familiar with the sequence of the planets. Their "nearness" to earth is based on which planets can eclipse or come in front of other planets.

Notice that the sequence can be resolved into three opposing polarities shown by the arrows, with the sun as the harmonizing middle.

First polarity: Saturn - Moon, Sun as middle
Second polarity: Jupiter - Mercury, Sun as middle
Third polarity: Mars - Venus, Sun as middle

Saturn -- Moon

Saturn is the most distant planet, just as the Moon is the closest. Saturn is like a gateway toward the spiritually active stars. The Moon leads to the ether spheres that are close to the Earth, where the Spirit wants to impress its seal into the world of substance. Thus, Saturn/Moon bestow the direction and strength for beings to incarnate. It is important during embryonic development and for the first thirty years of a person's life. In man, the Saturn process enters the hair vortex at the back of the head and radiates into the body, making Man into a picture of his individual Ego. This process terminates in the skeleton, where it becomes a lifeless image of the Ego. If Saturn were to work by itself, in thirty years man's body would have calcified and turned into a stalactite!

Saturn works as a differentiating force within warmth, the most spiritual ether. Where Saturn rarefies warmth, crystallized substance (bone) comes into place. Where it densifies, warmth's carrier blood (bone marrow) is formed. Blood cells are formed in the centre of the physical, almost lifeless skeleton. After about three weeks, blood cells are absorbed by the spleen. So the spleen is the consummation of the Saturn process.

We can see two Saturn processes:

1.   Incarnating: leading to a hard, lifeless image in physical space.

2.   Resurrecting: the Ego, in the form of the blood, permits karma to play out over time.

Saturn is both death and resurrection. The image of the Ego occurs twice, once as the physical in the skeleton, and once, through the blood, as the being living in time. Lievegoed summaries these two processes with symbols, ^ or v. Saturn leads the spiritual to the physical world, but in so doing, it brings dead rigidity. But Saturn leads the spiritual out of the physical in the course of one fulfilling one's life (karma), a form of resurrection. In the plant world, we usually meet with only the first process. As Saturn influence is coming in from the far outside, it can be effective only where working comprehensively from all sides, not in working from a centre. Saturn is active in the Valerian preparation, which envelops the whole outside of the compost pile. Such compost bestows full expression of the seed's species.

The two constellations ruled by Saturn are Aquarius and Capricorn. Aquarius character is defined by it being 'Fixed Air', thus its natives have strong opinions and are often philosophers and social renewal activators. Capricorn is defined as 'Cardinal Earth' and its natives are the corporate entrepreneur who build empires through sheer hard work and determination.

In this first paragraph we see Saturn as the spiritual archetype. The seed thought by which all else can form around. This can be easily associated with the image of Aquarius. Its Fixed Air nature provides an image of a strongly held thought. An ideal that manifestation forms around.

This primary Saturn process works right into the formation of the skeleton and if not checked would turn us into stalactites. Once Aquarius fixes on an idea they are not known for letting go or willingly adapting that vision. They are right and often end up 'grumpy old men' when their perfect ideal is unable to adapt to the needs of the time and thus fails to gain support of the community they crave.

The skeleton is often related to Capricorn, due to their focus on physical structure. However in this case, Capricorn's ability to build appropriate forms up over a long period of time, links it more closely to the secondary function of Saturn. The Cardinal aspect of Capricorn, provides the adaptability to life, inherent in the secondary function. This is the sign of the individual appearing to live out their karma, through their commitment to their task, regardless of many other aspects of their lives. They are driven individuals with a purpose. Clawing their way to the top of their mountain.

In this last paragraph we have an image of the preparation Valerian and the telling statement that we see it enacting the first / primary process of Saturn most. However when sprayed on plants Valerian encourages them to flower and seed more profusely.

Now consider the Moon - it stands opposite to Saturn and manifests continuity of qualities through a sequence of generations. Moon processes are active in propagation and heredity, in the sense of cellular reproduction. In this case, we observe cell rising from cell, linked in steady growth. The Moon is associated with recurring repetition of the same. She is concerned with the sequence of generations streaming horizontally across the Earth and continuing over time. Moon forces are active in swelling growth, in cell division and in propagation. If Moon forces alone were active in Man, he would become a soft sphere of albumen and would grow forever. But the Moon finds a limit in the skin; it is not effective outside the boundary.

Moon forces enter in front, in the region of the bladder, radiate into the reproductive organs and eventually permeate the whole organism as far as the skin. While Saturn is the bearer of individuality to the skeleton, Moon is the bearer of the type and becomes evident in the skin. Humans with strong Moon forces have beautiful skin (note how we associate the beautiful skin of movie stars with a strong sexual attraction.) The skin as "Moon skeleton" gives an image of Man's heredity.

Moon processes are also differentiated. During embryonic development, the nervous system forms out of an invaginated strip of skin, placed within like a little island. This "inner skin" becomes the bearer of a second process -- that of the   external world reflected through the nerve/sense system until it reaches the human's consciousness. Thus, the brain becomes where the Moon forces come to a rest. Physiologically, the force now holds back mere cell division and the brain tissue becomes the most differentiated in the body. Such tissue also suffers the most loss of connection with etheric life forces. Ironically, in giving us the ability of creative thought, the brain tissue becomes the least "alive", the least able to heal itself or reproduce. This end result of differentiation is called intensification.

So the Moon also shows two aspects:

> 1.Growth promotion: streaming in time, eternal repetition of the same
> type.

> 2. Repressing life: mirror-like reflection, life-forces bounce away, but the image of the outer world is preserved. (Note how the Moon shines by reflecting the Sun's light and that silver is used to make a mirror.

In the plant world, we see Moon effects in cell growth. Calcium acts with the   watery element as mediator of this growth. The Oak Bark preparation, containing high calcium, is prepared in the animal skull (as if a brain) and held under water. Steiner said this preparation would counteract plant diseases by suppressing an overly rampant etheric element. One could also say, by setting a limit (a skin) to the Moon forces.

As we look as these processes weaving into one another, we see Saturn as the image of the Ego individuality, placed in space but becoming another opportunity in time. Meanwhile Moon places into the stream of time a sort of type-heredity that is not individualized. The human overcomes the sequence of generations by repressing blind growth and, in reflecting the external world, the mind awakens.

When it comes to the Moon, the constellational references are somewhat easier to find. The primary processes of the Moon are describes as endless cell division that would go on forever. In Eugene Kolisko's booklet on the "The Twelve Groups of Animals" he described Cancer as ruling the Protozoa. This is the 'original family' of single celled life forms from which all else developed. Cancer is traditionally ruled by the Moon and is considered the constellation of the nurturing mother, through to the tribe, that supports the families survival, and thus it is the basis of life and its development. It is therefore not difficult to find   Cancer as the constellation of the Primary Moon forces.

The secondary process is not as clear an association, however we could allot Leo to Moon 2, due to the clear association of Cancer to Moon 1, and therefore Leo has to be Moon 2.

However in the secondary process the endless process of undifferentiated cell division is 'matured' and directed into forming specific tissue. It reaches its extreme in the brain. So extreme, and complete, that the etheric forces the moon so strongly carries, are nearly completely consumed to the point that nerve cells do not readily regenerate.

The brain reaches its pinnacle of evolution in the mammals. Kolisko identifies Leo as the ruling constellation of the mammals. Leos 'lower' qualities, as imaged in the tropical, Aries > Pisces zodiac, are reflected in its rulership of the fifth house, the house of creativity and artistic endeavor. Art can be said to be an attempt to reflect an image of something, that can not be said in any other way. It is a non verbal "reflection of an image of the outer world preserved." Thus fitting the image of the Moon's secondary process.

Jupiter - Mercury

Jupiter is the moulder of the world. While Saturn creates a bare image in the skeleton, Jupiter moulds the rest of the body around the skeleton in fullness and flowing beauty. Jupiter works down from above, rounding and recreating the celestial sphere in rounded (plastic) forms. All the internal organs are rounded, sometimes with hollows where another organ's roundness presses in. Jupiter forces radiated from the rounded brow, moulding the brain, and later the thoughts, especially those concerned with bringing order and universal connections. From there, these forces work deeply into the body, forming organs and muscles. At the same time, Jupiter strives for super-human gestures. If Jupiter worked all alone, at about fourteen years old, we would all be like beautiful Greek statues, with soulful bearing and gesture but completely rigid.

From this rigidity, the Ego individuality frees itself in movement and gesture. The alternating play of muscles contracting is deeply connected with the liver. Muscle movement is due to chemical changes in the tissue, but these chemicals must be controlled and balanced by the liver. This organ is the most fluid and life-filled, that is, it is the organ least permeated by Jupiter processes so that it can have the role of controlling those forces.

In the plant world, we see Jupiter forces creating the rounded moulded forms and secondarily in chemical changes, where Jupiter and Mercury work together in the flowing sap. Jupiter is represented by the Dandelion preparation. Moulding forces are handed into the plant through silica, the mineral associated with forces from beyond the Sun. Rudolf Steiner described dandelion as the mediator between the cosmic silica forces and those surrounding the plant, bringing health and stability.

With Jupiter we have two principles identified, which can be summarised as 1) Plastic forces which mould around the Saturnian 1 archetypes, and 2) the   Pharmacology of living forms.

The two constellations ruled by Jupiter are the Water / Mutable Pisces, and the Fire / Mutable Sagittarius.

Pisces, the mutable watery constellation, is seen as the constellation that easily moulds and adapts around the rules of the game, to find the most appropriate solution to a circumstance. Its reality and response continually changes according to its environment it is in. Due to its relationship with Neptune it is considered the chameleon or shape changer of the zodiac.

Sagittarius, with its fiery concern for universal truth, that associates it often with the image of the alchemist, suggests it is the principle with the energy to fuel the warmth processes necessary for "Pharmacology" to occur.

In this regard there is an interesting set of correspondences worth considering. Dr Steiner says (21) ' In pursuance of these phenomena, we find we must study on the whole these two polarities throughout the organism, these radiations and that which opposes them and dams them up. There is need to keep this distinction in mind, for all that tends to form albumen, in the manner described above, is associated with the damming up action, and all of a metallic nature introduced into our bodies, has to do with the radiating forces." and a little later " We must attribute much of the result achieved to two substances, which receive far too little attention in their effects within the human organism: these are fluorine and   magnesium. In the — so to speak — rarefied form in which they occur within us, both fluorine and magnesium play prominent parts, especially in the process of shape formation in the child, up to the change of teeth. The forming and fitting of the solid framework in the human organism takes place through continuous interaction between the forces of magnesium and fluorine respectively; in this interplay, the forces of fluorine act plastically, mould as a sculptor moulds, fill out contours and bar the way to the forces of radiation, whilst magnesium acts as a radiating force and constitutes the fibers of tissue, etc., into and along which the substance arranges itself. It is not a senseless phrase, but wholly in accord with the course of nature to say that a tooth is formed thus: It is shaped, as far as its circumference and its cement is concerned, by the activity of the plastic artist "fluorine," and magnesium pours into it the forces which have to be shaped to a plastic form."

These paragraphs highlight a basic principle we meet throughout out the study of Biodynamic Chemistry. That being an active polarity between radiation and conclusion or as RS says 'damming up'. In this context RS is talking of the teeth and the halogen, Fluorine and metal, Magnesium. This is but one example of the principle seen continually through life processes. We can broaden these images though and see that all Halogens have this damming up effect, while all metals have the radiating effect. We will see later in my exploration of the Cell Salts, how this same polarity, between radiation and conclusion, can be found within the   Anions.

What I wish to establish with this quote is, that the plastic forming process Lievegoed describes, are a Jupiter process which RS clarifies as a Halogen process.

Within the context of Jupiter's rulership, it is worthwhile further referencing Dr Hauschka in his 'Nature of Substance', where he talks of the Halogens. He talks of the polarity between the Halogens (F, Cl) and the Alkalis (Na, K). The alkalis he identifies as, forming closed sheaths and colloids which work as up building forces, that sustain and promote life processes. While acids and the Halogens in particular, are "dry, contractive and hostile to growth", they 'breakdown, burn and dissolve", 'they press towards decisive action; they either curdle colloids or reduce them to a true solution". "Hydrofluoric acid can melt the end of a glass rod" just as the jagged teeth are rounded off by enamel as they emerge". He goes on to use an example of an illness where body parts are not 'rounded off' or concluded, eg fingers, nose chin, toes, and that this is due to a weak halogen process. He goes on further to say the ancients saw these processes proceeding from the constellation of Pisces. " Just as Virgo symbolises the selfless offering of an enclosed sheath within which life develops, Pisces pictures an active coming to grips with the world and destiny.

Furthering the example of the teeth, Hauschka relates Magnesium to Sagittarius. Of Magnesium, Hauschka summaries its dual action, as a hardening agent that compresses life into solid earthly form and on the other side, it activates light forces. In this sense we see an image of the secondary Jupiter process. Mg works as the light bringer, that fuels the chemical reactions that fuel plant growth in the leaf region, and plant growth in general. In humans, Mg allows forces that previously served organic functions, to be set free in the form of a capacity to think and remember. An activity that can easily be associated with Sagittarius, as the philosopher, teacher explorer.

The Jupiter organ, the liver, also has two distinct processes it performs. It selflessly cleanses the blood and it metabolises all manner of substances for the bodies chemistry to function correctly. In this dual activity we again see a selfless process and a (al) chemically upbuilding process. The selfless process, being associated with the 'humanitarian' Pisces, cleaning the blood, and the alchemical nutritive process being associated with Sagittarius.

While Jupiter gives harmony and order, Mercury brings chaos, but one we might call "sensitive chaos". This is movement without any particular direction, but ready to flow into anything, adapting to any resistance by flowing around, always in movement. One has only to think of liquid quicksilver for an image. Whatever kind of movement happens, it is in response to outer circumstances. In the human body, this type of streaming movement is represented by the lymphatic system. The blood vessels may be fixed but lymphatic streams can move as they chose, as long as they reach the lymph glands. While Jupiter is harmonious symmetry, Mercury has a tendency to asymmetry. If a feature is oblique or crooked in the human face, that is Mercury confounding Jupiter's intentions. Mercury has a sense of humour and is pleased if divine intentions don't quite come off. Thus, the gods never finish their work because a state of flux remains. If Jupiter is the king sitting on his throne, at his feet is Mercury as the court jester. Mercury tolerates all conditions, heat and cold, sun and shade, but wants life to go on. In extreme cases, Mercury becomes dishonest and even leads the plant into parasitism. Mercury was the god of merchants and thieves, making sure that earthly goods do not remain in one place but are constantly changing hands.

This faculty for adaptation would lead to loss of character; Ego avoids this by diverting movement with more movement. We see when two streams meet, they form whirlpools and empty spaces. This is Mercury's second principle - creating a responsive "newness" by combining movements. The organs taking shape from dynamic movement become different from the archetypical images being projected to Earth. In the plant world, we see these two principles. A beech leaf is completely different from an oak leaf - variability is a Mercury expression. Merging forces are healing, but true healing requires not only that forces merge but also that something new results from the merger. Mercury is active in the Chamomile preparation. This preparation stimulates plant growth through potassium and calcium and has been intensified by the intestine.

Jupiter and Mercury weave into each other; the preordained form of the organs meets Mercury and is changed according to the circumstances. The chemical expansion and contraction of Jupiter give direction to the streaming movement of substance through Mercury. Jupiter comes to rest in muscle, then the activity changes to chemical and the resulting muscle movement overcomes what would otherwise be rigidity. Muscle movement sucks at the liver. (That is, the muscles fetch substances from the liver rather than the liver sending to the muscles.) Mercury streams in an irregular way through the lymph vessels and rests in the lymph glands, where the fluid stream leaves the body. The one exception is the lung, which as an empty hollow lacks the fluid streams present throughout the rest of the body. Thus, the liver and lung are the consummation of Jupiter and Mercury respectively.

The two processes of Mercury are characterised as the primary process being "Mercury brings chaos, but one we might call "sensitive chaos". This is movement without any particular direction, but ready to flow into anything, adapting to any resistance by flowing around, always in movement."

While the secondary process is " creating a responsive "newness" by combining movements. The organs taking shape from dynamic movement become different from the archetypical images being projected to Earth.

The two zodiacal constellations we have to consider are Gemini and Virgo.

Gemini is considered a 'youthful' sign that constantly moves between its two twins. Psychologically, this is generally experienced as an extrovert, outer social, and inner introverted duality of personality. Being the Air /Mutable sign , it is characterised as being constantly moving and is the zodiacal 'butterfly', that knows a little about a lot, but not known for its depth or lasting achievements. Hence it is easy to associate the Mercury 1 process to Gemini.

 Hauschka associates Sulphur to Gemini. Sulphur acts as a catalyst in the interactions of the 'big four' elements of protein, H, N, O & C. It is the 'oil' that   allows the spiritual bodies these elements carry, to interact with each other. When it is insufficient, it leads to a 'sticking' of the bodies, which manifests in Humans as the autism spectrum of aliments. Where there is too much S, it leads to a 'slippery' relationship between the bodies, resulting in the hysteria spectrum of illnesses. Similar results can also be observed in plant growth, however these   appear as stunted and reduced growth when there is a lack. Sulphur is therefore the element of movement. In its multitude of relationships, it is THE prime unquestioning facilitator, of biochemistry.

Virgo on the other hand is Mutable / Earth and the sign of the craftsman. In Greek myth Virgo is associated with Hyphastis or the roman Vulcan, who acted as the maker of all manifest things. Anything the gods used, including the humans, was made by Vulcan.

Lievegoed's comment "We see when two streams meet, they form whirlpools and empty spaces" is very indicative. Whirlpools, are vortexes, which act as the basis for the organisation of forces and matter, within creation. In the middle, we find the 'empty space', manifesting as a vacuum. RS has commented that if one was to view matter from the spirit sphere, one sees only a black hollow vacuum, where the object stands. As, from the spirit view, one sees the organised spiritual activity, that holds the vacuum in place. Matter is sucked into a vacuum, and so once the spiritual forces are organised, manifestation must take place. In my chapters on the planets, (22) I outline a 12 fold planetary pattern, that places Vulcan with Virgo as the 11th planet. A planet that governs direct manifestation of matter from spiritual imagination.

Hauschka's association for Virgo is the Alkalis. He describes the alkalis, (Na, K) as creating membranes and colloids, within which life processes can be sustained. Virgo symbolises the selfless offering of an enclosed sheath within which life develops ". This supports the image of Vulcan as the creator of life as needed at the time, and Mercury 2 being ruled by Virgo.

Mars - Venus

Mars, the last of the outer planets, carries purposeful movement that is directed towards a goal. Mars-force is the process by which the archetype penetrates into the physical sphere but also that which pushes it out again into the world. Mars is at work as the growing point of a plant pushes out and conquers space. Mars brings an inner activity of determination and direction - without Mars, no plant would exist. Mars forces work in the shooting and sprouting of spring growth - as an athlete throws his javelin.

Mars forces enter Man between the shoulder blades and penetrate into the iron processes of the blood. They also radiate into the speech centers; Mars is formed in the words streaming from the human mouth. The Mars type of human is outwardly active all the time but finds himself unable to preserve what he has   created because he can't stand to finish anything. Rather than care and cultivate, the Mars type destroys and builds anew. He has a strong creative urge but, if blocked, becomes consumed with wrath.

Mars does not yield to soft measures, so if the Ego is to resist being carried away, it must summon strong resistance. Such opposition creates a damming up of the direction but surprisingly, this transforms into sound. Think of the string on a musical instrument. Force and resistance interact as tautness - and the string forms a musical tone. If we take a metal plate, cover with fine sand and strike a tone, we observe the sand particles form "sound figures" indicative of the particular tone. On Earth, substances come into existence according to Tone Ether forms. Cosmic music starts from Mars and is passed on to the Earth through the Chemical or Tone Ether. In the human or animal body, Mars orders and forms the substances working from the astral body through the organs. In the plant, there is no astral body and Mars works from the cosmic sphere.

Iron forces are active in blood haemoglobin and end in the liver where the haemoglobin molecule is dissociated into gall. Iron is held back within the body. Out of this holding back are born albumen-forming forces like "sound   figures". The process of building protein in the liver is one of arranging the chemicals like sound figures. Mars is manifested in the Nettle preparation, which harmonizes forces in the soil, bringing true nutritive value to the plant. This is connected with a health protein-building process and also the formation of starch, because each starch granule is surrounded with a protein sheath.

The two constellations associated with Mars are The Cardinal , Fire Aries and Fixed, Water Scorpio.

The two processes of Mars are described as 1) carries purposeful movement that is directed towards a goal. Mars-force is the process by which the archetype penetrates into the physical sphere but also 2) that which pushes it out again into the world. Mars is at work as the growing point of a plant pushes out and conquers space.

The Mars type of human is outwardly active all the time but finds himself unable to preserve what he has created because he can't stand to finish anything. Rather than care and cultivate, the Mars type destroys and builds anew.

This last paragraph could have come straight from an Astrology book, as the description of the Aries individual. Aries the ram, is renowned for rushing in head first, and thinking about the outcome later. There is always some loss somewhere with Aries. Sadly it is often others, and not the Aries that experience that loss.

Hauschka has allocated Silica to Aries. He adds "the silica process shapes life out of the cosmos as a sculptors hands shape clay" " Macrocosmic ideas underlying outer forms of life." "The Sun which is the mediator of these forces transits the constellation of the Ram in April, just that time of year when nature rises to bring forward a wealth of new forms. The seasons when nature rises to bring forth a wealth of new forms." " The world of appearances selflessly receives the imprint of archetypal patterns from creative heights in the symbol of the backward looking Ram".

In these references we can see the 'pushing into space' image of the Mars 1   process.

The secondary Mars process is described as-
Such opposition creates a damming up of the direction, but surprisingly, this transforms into sound.

Mars orders and forms the substances working from the astral body through the organs. In the plant, there is no astral body and Mars works from the cosmic sphere.

Out of this holding back are born albumen-forming forces like "sound   figures".

The secondary Mars process is concerned with the building up of protein. This is a transformative process of lifting of carbohydrates, through its interaction with nitrogen and the astral body, into a new substance protein.

While proteins can be found in plants, it is more characteristic of the animal   kingdom. So this secondary Mars process is the transformative process that lifts the plant kingdom into the animal kingdom. The legume and Solanace families are the two most common plant families which concentrate nitrogen alkaloids. The legumes form protein while the Solanace produce nitrogen poisons.

Scorpio, through its association with the phoenix image, is classically considered the sign of transmutation of elements. It rulership of the eighth house, makes this the house of emotional and sexual alchemy. This is place where, after the partnership is formed in Libra, the mutual attributes and resources of both partners, are fused together to provide the combined base for the future outcome. It is the house where individuals meet the spiritual influences of creation, albeit still unconsciously, and are transformed by them. This leads to more formal education, in Sagittarius's ninth house, to bring consciousness into the deep life changing experience, had in the eight house journey. This is the first stage of the spiritual journey which can unfold through the rest of the zodiacal houses. This is where the astral body enters to fertilise the occult union two individuals which   results in procreation, in all its forms.

Hauschka has Carbon as the chemical element of Scorpio, but notes Carbon has a very special relationship with Iron, the metal of Mars, and that through their combination both are transformed. Carbon changes iron into steel and iron changes carbon into diamonds. It is this transformative ability of Carbon to be the most basic framework of all life, yet it reaches such sublime heights in the diamond, that he sees it emphasises the motif of the transformative Scorpio.

While Mars is out there, Venus is goes deep and stays hidden. Venus is connected with deeper nutrition (at the cell level) and the deepest building process of the organism where inanimate substance is first received into the life stream. Venus is connected with clearing space for something else to unfold. Think of the home belonging to a quite and attentive host. In this home, people can meet and have a productive exchange of ideas precisely because the quiet host has arranged the conducive environment. As Mars is associated with speaking, so Venus is associated with listening. Goethe called conversation more precious than light because of this harmony between partners.

To become Venus completely would mean to renounce the self; the Ego could no longer exist. Ego overcomes this by "sucking away" or excreting products of the life forces, through the kidney-bladder system. From these organs, the sucking power of excretion reaches into all the living cells. The Venus process comes to an end in the kidney. Here ether force and substance are separated, the substance is excreted and the ether forces radiate upwards into the eye, combining the act of seeing with the force of going out into the world. Mars and Venus have a strong cooperation. Protein, formed through Mars, nourishes the cells along the Venus path. The ether forces freed by the kidney unite with the determinations and   direction of Mars in the eye's force of vision.

Think of the violin. The determined movement of the bow is halted by the stable string and tone results. The resonance of the instrument creates the opening for the tone to develop and individuality adds musical quality to the tone. In the plant world, we can see the force of the growing point, directed outwards, is surrounded and filled by nourishing Venus force. Both are needed for growth. Building up and excreting cooperate closely; the dammed force of the growing shoot makes protein formation, while the force of excretion is in the bark and its deposits. Venus is manifest in the Yarrow preparation, which is prepared in the stag bladder, at the end of the kidney sucking process. Thus, the yarrow is strengthened in its connection with Venus so it can receive cosmic substance to enliven the soil and balance exploitation. It is associated with potassium processes.

The two constellations we have to define here are the Fixed Earth, Taurus and the Cardinal Air, Libra. Taurus is the constellation in Greek myth associated with Demeter— the goddess of fertility and agriculture. Libra is associated with her daughter Persephone, whose claim to fame was being abducted by Hades and taken into his underworld domain, where she became his very willing partner.

Venus primary activity is described as "Venus is connected with deeper nutrition (at the cell level) and the deepest building process of the organism where inanimate substance is first received into the life stream." Lievegoed talks of Venus opening up the etheric body, so that the astral can be received. It is here we see the images of the gracious host making the place for a nurturing environment to be created. Demeter as the nurturing goddess of nature, which provides the environment for all life to take place, is the ideal image of this process. We only need to remember when Demeter's daughter, Persephone was taken from her, that she withdraw her life giving forces, and all life was threatened with extinction. Taurus folk are well known for their love of food and sensual pleasure, thus easily fitting Taurus with this image of nurture.

Hauschka's association for Taurus is Nitrogen and he associates nitrogen penchance for movement with Taurus. Given Taurus is a fixed Earth sign I do not see this relationship.

Venus' secondary processes is "To become Venus completely would mean to renounce the self; the Ego could no longer exist. Ego overcomes this by "sucking away" or excreting products of the life forces, through the kidney-bladder system." Venus, as the perfect hostess does not only provide the nurturing environment, she also cleans up afterwards. This is imaged in this part of the processes relationship to the kidneys. Libra is the traditional ruler of the kidneys. Any affliction to Libra in a birth chart generally shows up in a weakness of the kidney and urinary system.

Hauschka associates the activities of Calcium with Libra, mostly through seasonal references, which as a dweller of the southern hemisphere, I have some difficulty with. However the 'sucking action' of calcium can be seen to be synonymous with that of the kidneys. It is Dr Koenig in his book 'Earth and Man', Lecture 5 on "The Kidneys", where a clearer association can be found. He states (pg 123) "the kidneys keep, not only the blood, but the whole fluid of our organism in balance. If there is too much they excrete more , if there is too little they excrete less." "The second function of the kidneys is the one whereby all minerals within the body, if they are in excess, are excreted. Again it is an act of balance. Sodium, potassium and so on, are either excreted or sent back into the blood stream." "The third function … The acids and alkalis within the body as well as the serum, are kept in equilibrium. If we eat too much protein, we excrete more acid substances; if we eat too many vegetables and become too alkaline, then we excrete more alkaline ions." It is in this constant balancing act of the kidneys that we see the activity of Libra—the scales. Judgements are continually being made as to what to keep and what to excrete.

Dr Hauschka, Dr Lievegoed, the Elements and the Zodiac

Following on from the considerations made with the zodiac and the planetary processes, it is relevant to include two diagrams with Dr Hauschka's chemical elements, the zodiac, as outlined in his "Nature of Substance". (2, Pg 130) and Dr Lievegoed's suggestions. And on through the Agriculture Course.

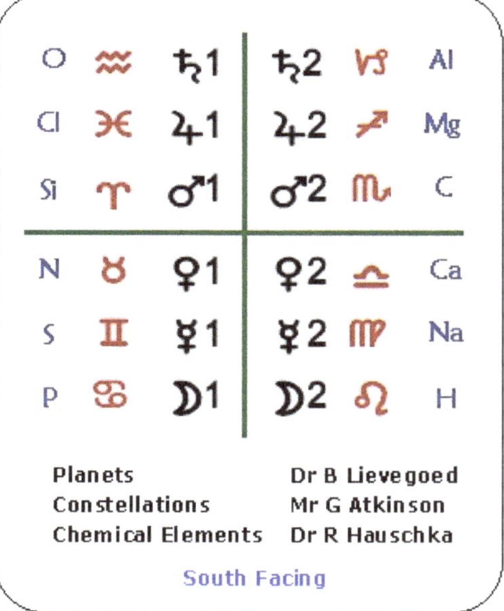

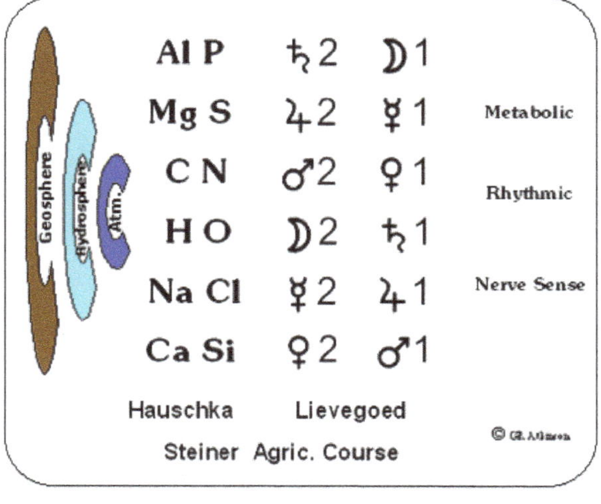

The Preparations and the Double Planets

Dr Lievegoed references the Biodynamic preparations to the double planetary processes, however he did not make specific instructions for how we might use the preparations to control either the incarnating or excarnating side of any planet. Initially I looked at which constellation related to which planetary process, and only got so far with this practical investigation.

In my experience of the preparations, I work with them as Force influencers. I use them to control the way the energetic activities work upon Nature. See the diagram on page 21. I see problems as an expression of the energetic activities interplay, and the homeopathic preparations influence the activities around the object, and the problems go away. Dr Steiner outlines this approach in lecture 6 when he talks of nematode control. This is also the basis for Dr Steiner's medical approach for Humans. All this has me putting the **BD preps on the right hand side of the Double Planets diagram, on the Force side.**

So what of the Manifest side of the activities? A few years ago I came across the little book by Dr Otto Wolff MD, called 'Remedies for Typical Diseases' In this book he talks of a series of 9 remedies RS gave called the **'Odoron Remedies'**. These remedies are all made from plant and minerals, and have very little potentising in their manufacture. They are very physical remedies. Upon investigation it became clear, they follow a very similar pattern to the BD preparations in their application to organ systems and their diseases, however in a much **more 'material'** way.

This lead me to proffer the following diagram. I have yet to do any experimenting with these remedies on plants.

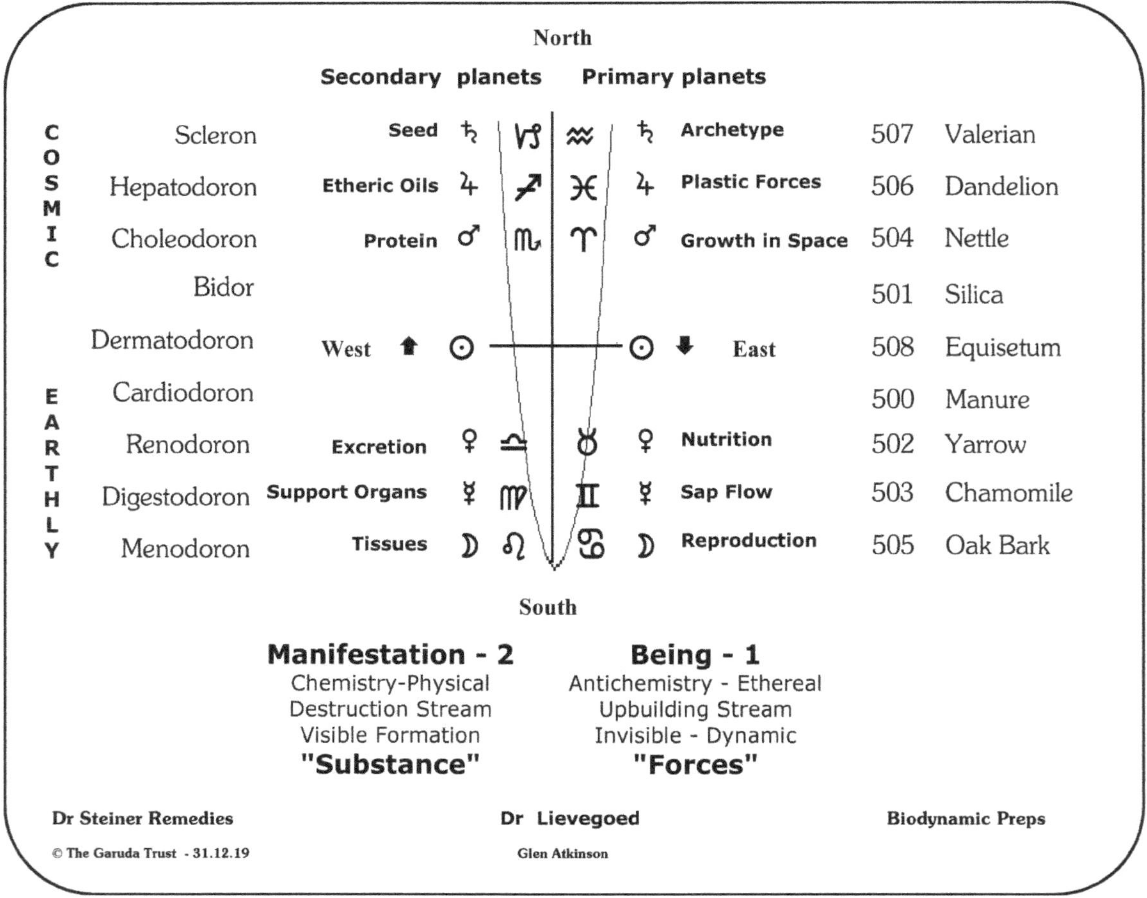

Stage 3

So far we have talked about the planets in the Agriculture Course, however this is not where RS story of plant growth ends. In other lectures he talks of the Earth and plants experience **through the seasons,** and how the Elemental beings help these processes along. This is where we move **from the Laws behind Manifestation to Manifestation itself**. We take a further step and the lemniscate enfolds upon itself to give us the **manifest physical plane**.

The overall story is told, in my essay RS Plant Growth story.   ' Cosmic Workings in Earth and Man', lecture 5, is where the 'Sap Stories' are told, and 'Man as Symphony', lecture 7, is where the Ethers and Elementals story is told. RS talked of this yearly cycle a few times in 1923. In April 23 he gave 'The Cycle of the Year' (23) and in early Oct 23 he gave 'The 4 Seasons and the Archangels' (24). Then the Agriculture course came in June 1924.

Each of these has there own language, and we are challenged to merge all his different pictures together.

When I did this, and further combined what I found from the medical lectures and chemistry, I came up with an organisation of the Cosmic and Earthly activities and thus the planets, that does not conform with the patterns found in the Agriculture Course. In this situation we can say that my end picture is wrong, or it is something different. The hint is that we are dealing with **manifestation ,** and using nature and its seasons as well as chemistry, as our reference. So this is a story of **formative processes IN manifestation**, and not formative processes BEHIND manifestation.

To come to my suggestion, that these other lectures are a Stage 3 step, I also referenced some other investigations. After several readings of most of RS medical lectures I came up with this image. I call this a 'Rosetta Stone' as this places around 10 different stories in relationship to each other. The one story that is altered from the Agriculture Course is where the Physical Formative Forces, and thus how the Double

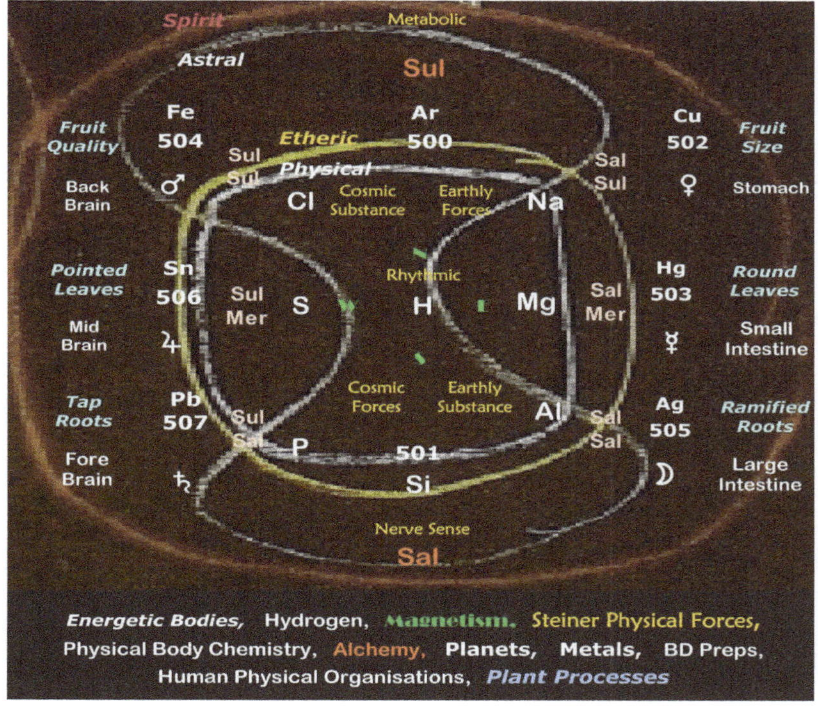

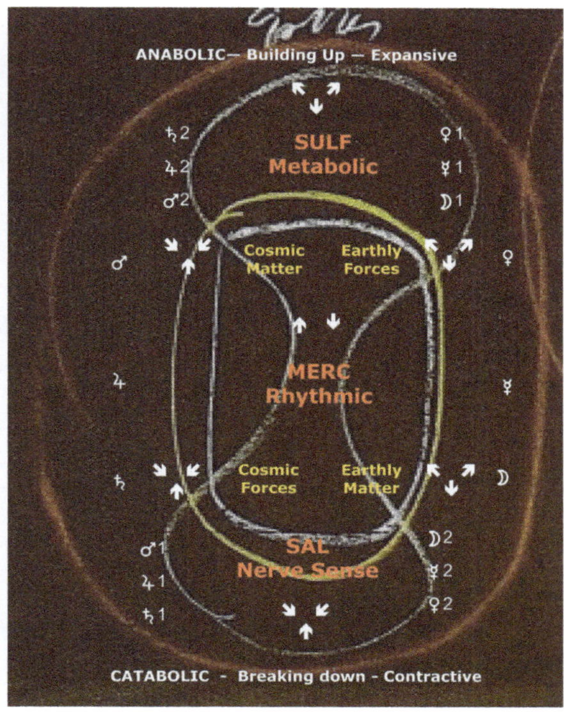

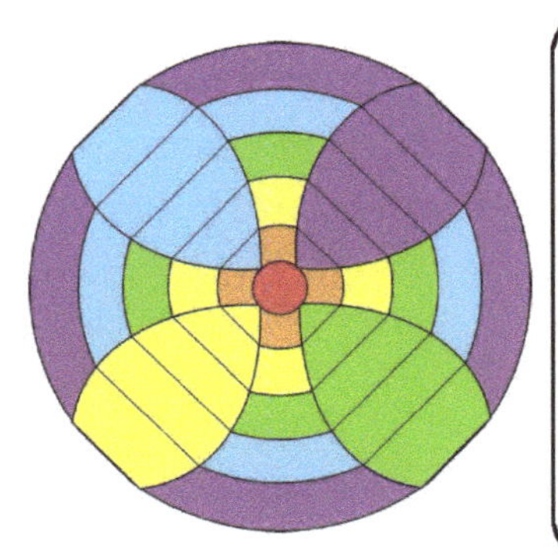

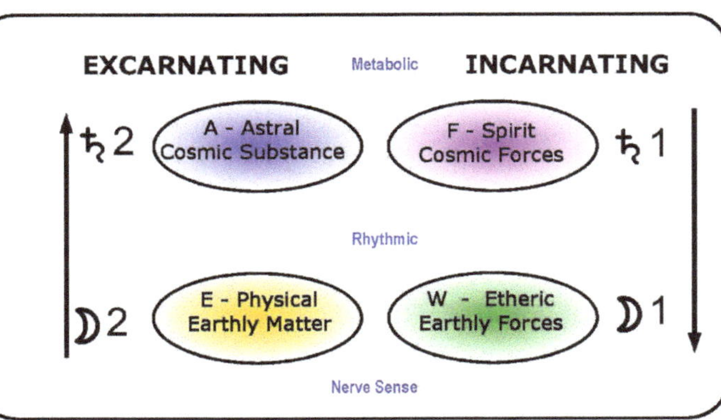

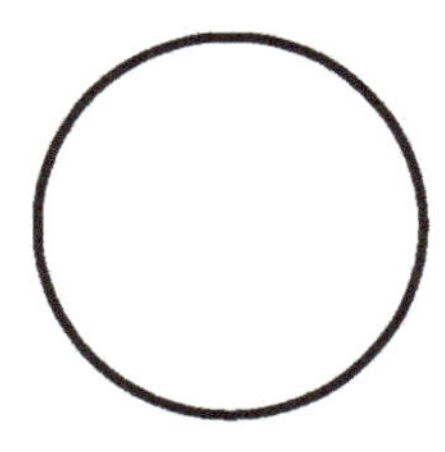

Lievegoed
Stage 1
ARCHETYPE

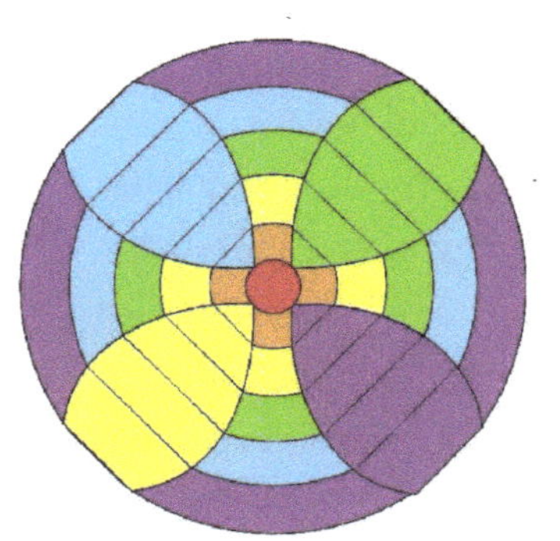

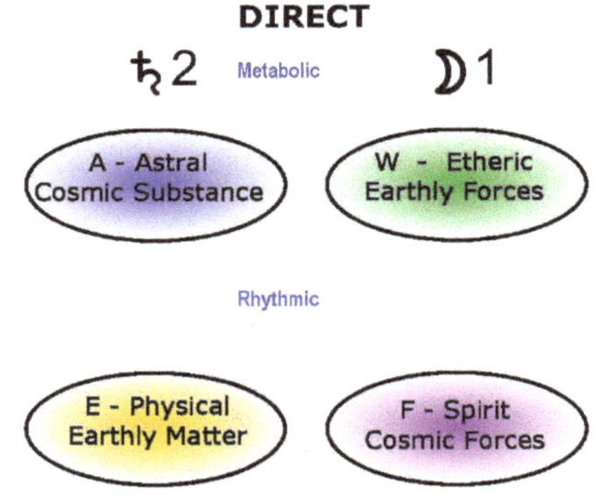

Lecture 8 Agric. Course
Stage 2
BEHIND MANIFEST

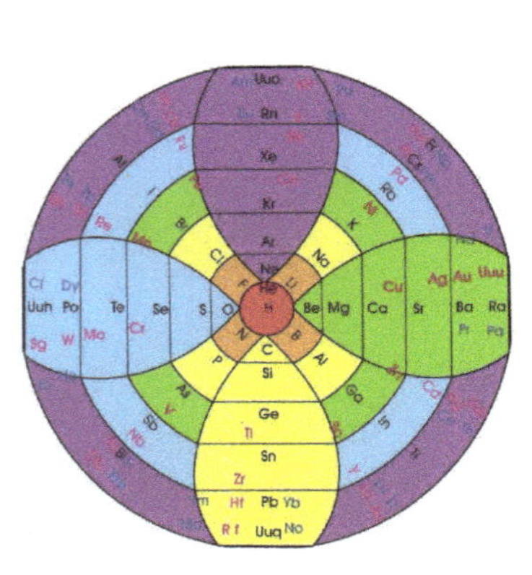

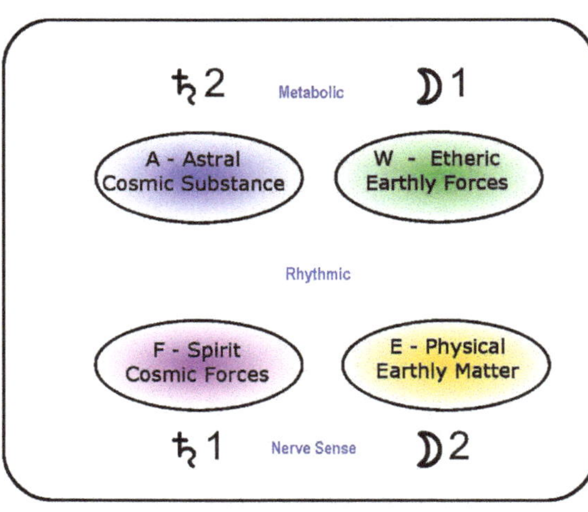

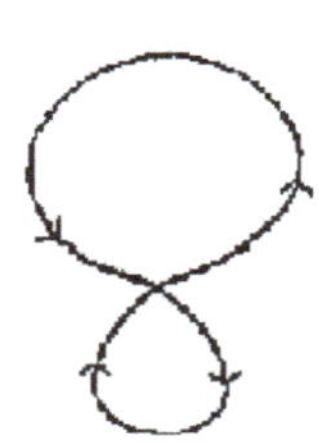

Seasonal Cycle
G. Rosetta Stone
Int. Phy PT
Stage 3
MANIFEST

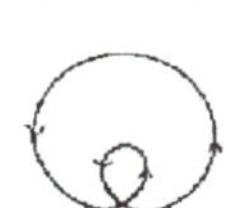

Planets sit, in relation to each other.

The Direct and Indirect planets stay the same, however a further flip of the Indirect group can be observed. The Cosmic Forces and Earthly Substance change places, providing us with an indication of the flow of the activities that dominate throughout the phases of the seasons.

Separate to this investigation I had earlier done an investigation into Chemistry. (1) My approach to Chemistry is a long story, and it took the better part of 20 years to move through all of the elements and their three groups into a Circular Periodic Table. This table describes the chemical elements as they appear in the External World. As we find them, and many already known associations and interactions, become obvious through this organisation. Which we can see is based upon the **World Physical Arm.** At a latter stage in this investigation I asked the question of **how could we see these elements work more deeply into Physical forms and even biochemistry**. The suggestion was to **spin the table 45 degrees** so that the table sat upon the **Internal Physical Arm**. What do we see when we do this? Essentially we find the chemical elements Steiner and Hauschka talk about in their medical lectures, are sitting in their appropriate place. The main elements they talk of are the 3rd ring or line of elements, which make up the basis of physical bodies. The elements of protein RS mentions in the Agriculture Course are one ring inside these elements. They are the basis upon which these next ring of elements work upon.

A further important finding was when I looked at the Metallic and Gaseous states of the elements, when placed on the Circle. This provides a very pictorial image of the polarisation of Metals and Gases, which conform with RS discussions in the 1920

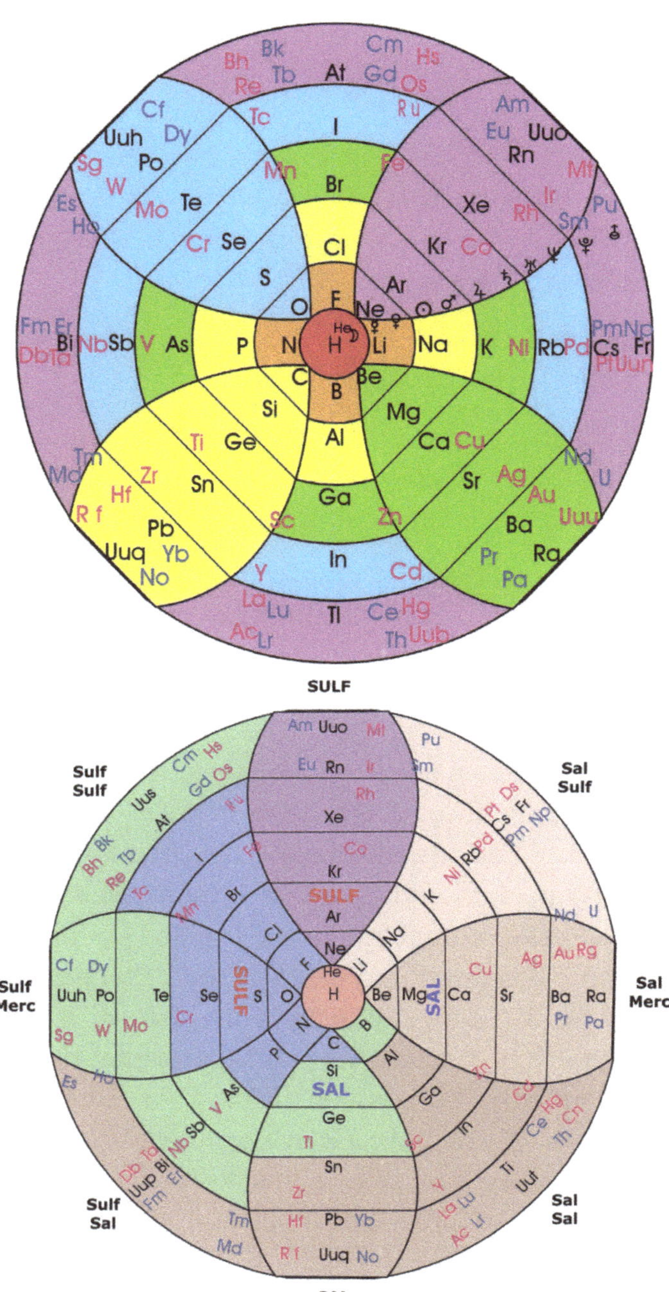

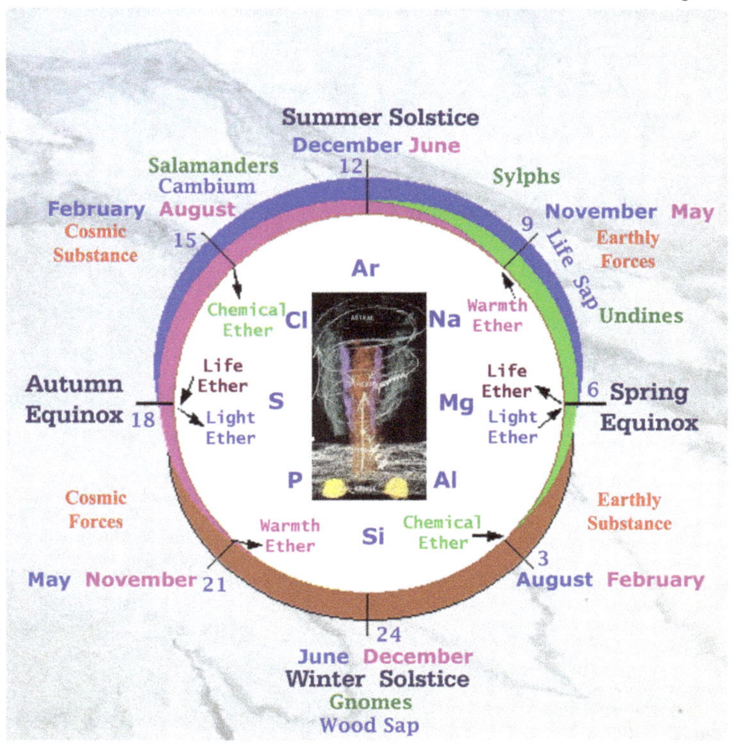

The Metallic States & Alchemical Chemistry

medical lectures , where he uses the language of Alchemy to describe the chemical elements interactions. This is manifestation in it hardest most real form, the physical states of Chemistry. The Brown elements are the metals, Green are the 'non metals' , Blue are the Gases, Purple are the Noble Gases.

We can observe the alchemical 3 fold division from the bottom, Metals Salt polarising with the Noble Gases, Sulphur pole.

Similarly, there is a clear polarity across the middle horizontal plane as well. The Calcium Metals Sal, to the Sulphur Gases Sulf. This provides a framework to look at the 'alchemical' nature of the elements, within a 3x3 form, and much more besides.

The next stage in this process is to place the images of the Sap Stories, The Ethers and Elementals, along with the flow through the seasons. This provides the diagram at the bottom of the previous page.

This order is there to see, and I suspect Dr Rudolf Hauschka, of Wala fame, has seen it.

One of the wonders of this image is, if we go four rings in from the outside, we reach the **Cosmic Physical Ring**. This is were Physical Bodies manifest. If we look at those elements we can see the polarity pairings, we have in Dr Hauschkas (RH) story. In the Nerve Sense Head Sal region, there is Aluminium and Phosphate, the middle Rhythmic system has Magnesium and Sulphur. And the Sulf poled Metabolic system has Sodium Chloride. Silica anchors this whole group in the Earth. The four elements of Protein, and RH's elements of the Atmosphere, are one circle in, with the 'Cosmic Elements'. Only **Calcium** is the outlier. It is one circle out, to be at the spot **where the Cosmic Etheric, Internal Etheric Arm and Positive Manifest Etheric Axis, meet**. It becomes a tag team with Magnesium, helping the Etheric to incarnate into the Physical Body. Calcium is the anchor of Life. All calcium on the planet has passed through some living being, to become the mineral we see. It is a manifestation of the Etheric, and anchors and stimulates the Etheric and with Magnesium's help anchors it into the Physical Body.   Stunning.

This ring of **Cosmic Physical** elements, work as an excellent model to place upon Dr Steiners medical lectures. We can also add a few of his 'Nature Stories', once we make the **flip of the Cosmic Forces and the Earthly Substance**. This is how they sit in the Circular Periodic Table, when based upon the Internal Physical Body Arm. This orientation provides very useful information about the internal processes of Life, as opposed to the information we gathered from the circle, orientated upon the World Physical Arm. Both orientations are effective simultaneously. This change of orientation, by spinning the circular PT 45 degrees mimicks the flip into manifestation in the earlier diagram for Stage 3. **A further step into matter .**

The 1920 medical lectures, talk of the Planets and the Metals along with examples of illnesses, and which of these 'Hauschka' chemical elements would solve them. Naturally the 7 fold Outer and Inner Planets make a good polaric framework to all this information.

The 1923 Sap and Elemental lectures tells of the growth processes through the seasons. Reading these I realised my previous PFF diagrams are all static. They tell of the processes and their interactions, but not of how and when Life is spinning. These two lectures do that. These are the stories of the PFF in motion, and how nature moves through the PFF activities, in the same order found in Chemistry and the Glenological Rosetta Stone.

Being at the manifest level, this is a good 'seasonal' diagram. Both planetary groups influences come directly from above, are accumulated in the Earth, before being released back upwards. While this process is a continual process, always taking place, there is also a seasonal component

B. Lievegoed

☽	Moon	♅	Uranus
☿	Mercury	♆	Neptune
♀	Venus	♇	Pluto
♂	Mars	♙	Persephone
♃	Jupiter	↓	Vulcan
♄	Saturn	☉	Sun

Steiners Agriculture Course

When we spin through the Seasons in the Saps and Ethers

The Planets in R. Steiner's Stories

Glen Atkinson

9.1.21

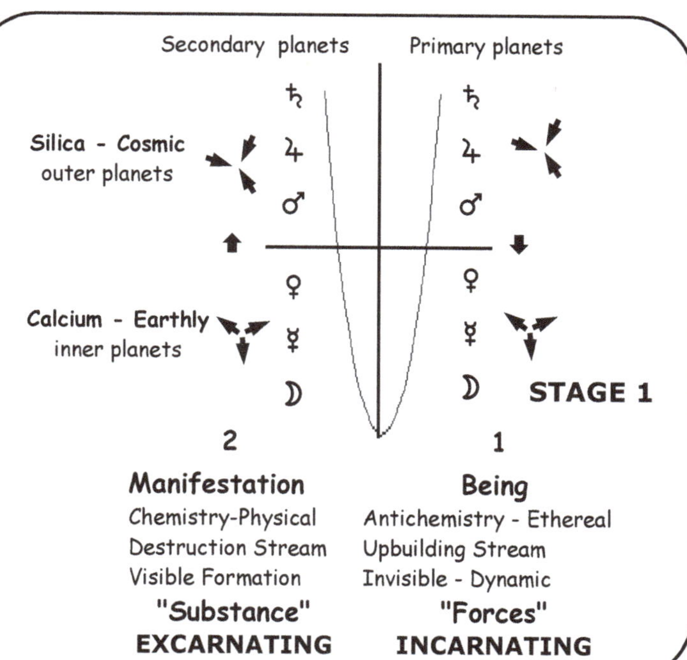

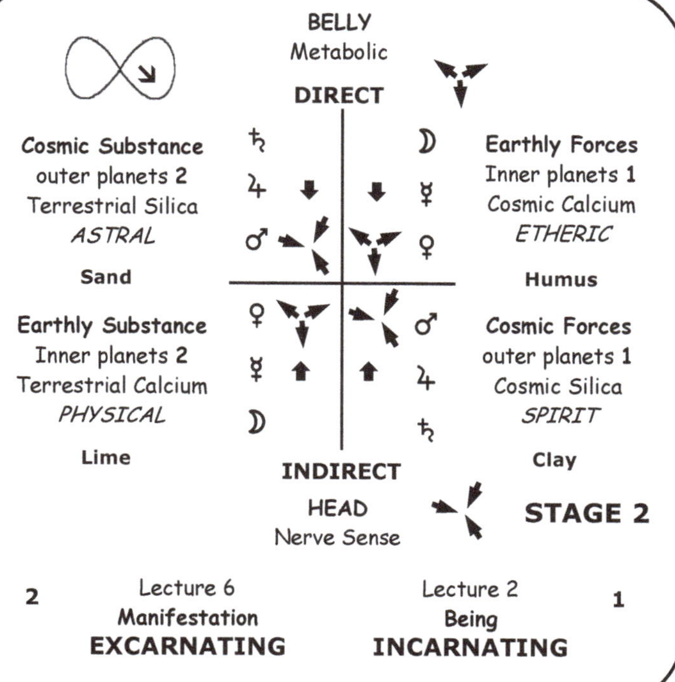

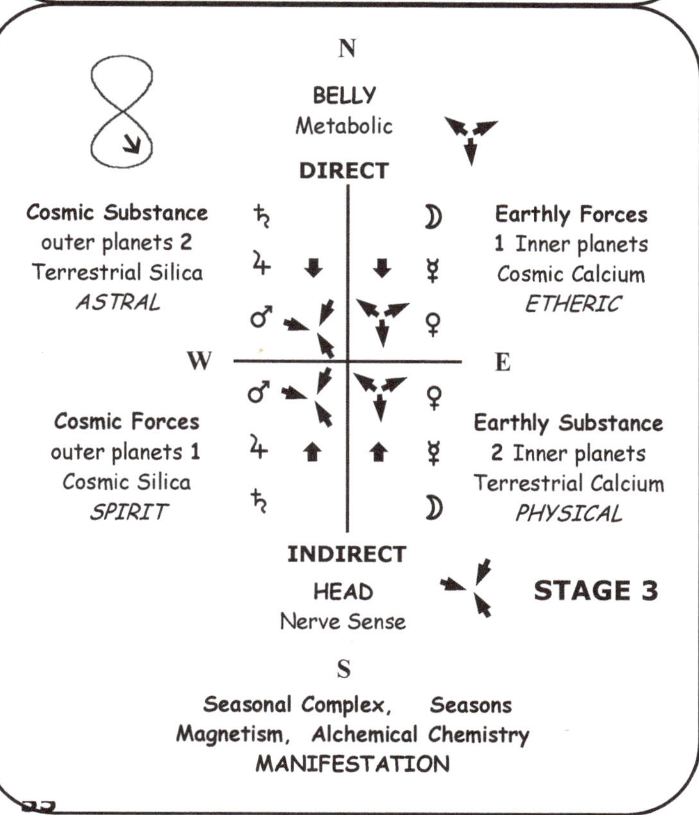

to it. As part of RS's 'Seasonal' story, the Direct Silica planetary influences are wound into the plant as both, growth processes, setting fruit and ripening it, while building the 'ideal plant form'; and held in the seed via the seed chaos process, as a plant spiritual imprint. Both these silica streams are drawn in through the Autumn, fructified and 'crystallised' around mid winter, and then released back upwards, as Indirect forces in the Springtime. This transformation is made possible by the Gnomes, who are the soil Beings. I am ok with it just being a natural 'organic' process, but in that story , that's the language.

The Calcium processes begins with sea life and animals metabolising Calcium into bone and shell, which is deposited into the Earth. This is then received by the plants and is carried upwards, with the Silica stream, swirled and potentised

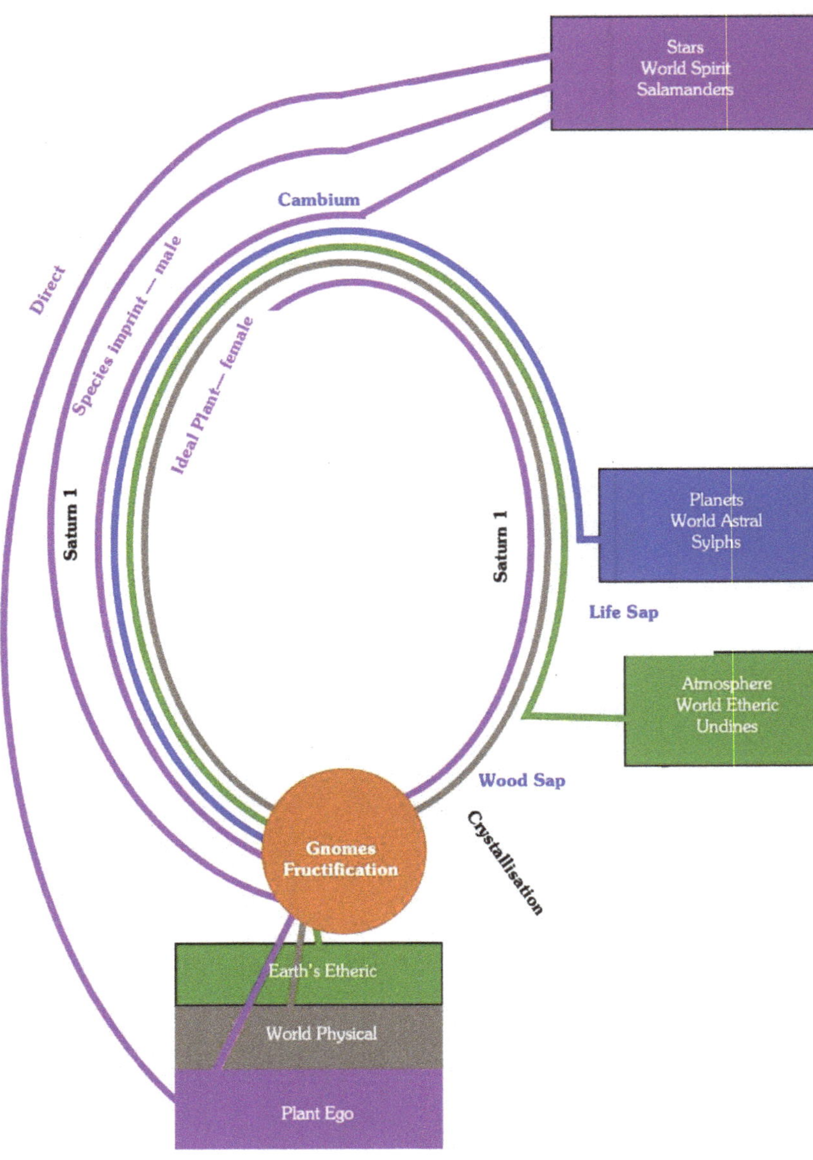

through the plant, into the transpiration flow, and into the Atmospheric moisture. The plants use this atmospheric moisture with light, in photosynthesis to make new plant material. All of this returns to Earth, as homeopathic Calcium, in the rain and dew, and decaying plant material, to be accumulated in the living Clay/Humus colloids within the soil, by the gnomes. This Calcium stream joins with the Silica stream, mentioned above, during winters fructification and begins its journey back through the plant, as 'Wood Sap'.

These two process cycles, are always interacting. RS tells of how this flow through the seasons, works with the various forces as it moves through. Our opportunity as RS students to use this as a guide for our observation of Nature and then as a basis for our actions with Nature.

Stage 1 is Lievegoed, Stage 2 is the Agriculture Course and Stage 3 are the 1923 lectures. None of these patterns are 'wrong' , they are just telling different parts of the story. There is a pathway through this maze.

Given all this 'evidence' I suggest we take this phenomena of the **3 Stages of Manifestation** seriously, especially when working with the planetary order, and the Physical Formative Forces. **These are the levers we can control nature with.** However we must know where we are, and the rules that apply where, in this game of three dimensional 'chess', that RS provided us with.

What does this all mean ?

Over the years I have made several attempts to show how this approach to the planets impacts upon the practical suggestions RS gives throughout the course. I have written a commentary to the Course , and done a reedit, putting relevant paragraphs together, while adding words to emphasis reading the course through this filter. I have also written some focused articles on natures cycle, along with ones on Fungus and Pest control. Please refer to these for more details. For the quick version I can show some of the larger impacting applications.

Lecture 2

Look at it again within this image.

Essentially it is a lecture outlining how the 4 astronomical activities manifest within the Physical body as the Physical formative Forces, within the threefold physical organism.

We are then given the four substances through which these forces work and through which we can control them.

Along the way we are given part of the seasonal story. RS talks of the mid winter crystallisation period, where the 'male' Cosmic seed chaos activity at pollination, joins with the 'female' Earthly Substances in the Fructification event, creating the Wood Sap activity that springs forth from mid winter, with the help of Clay.

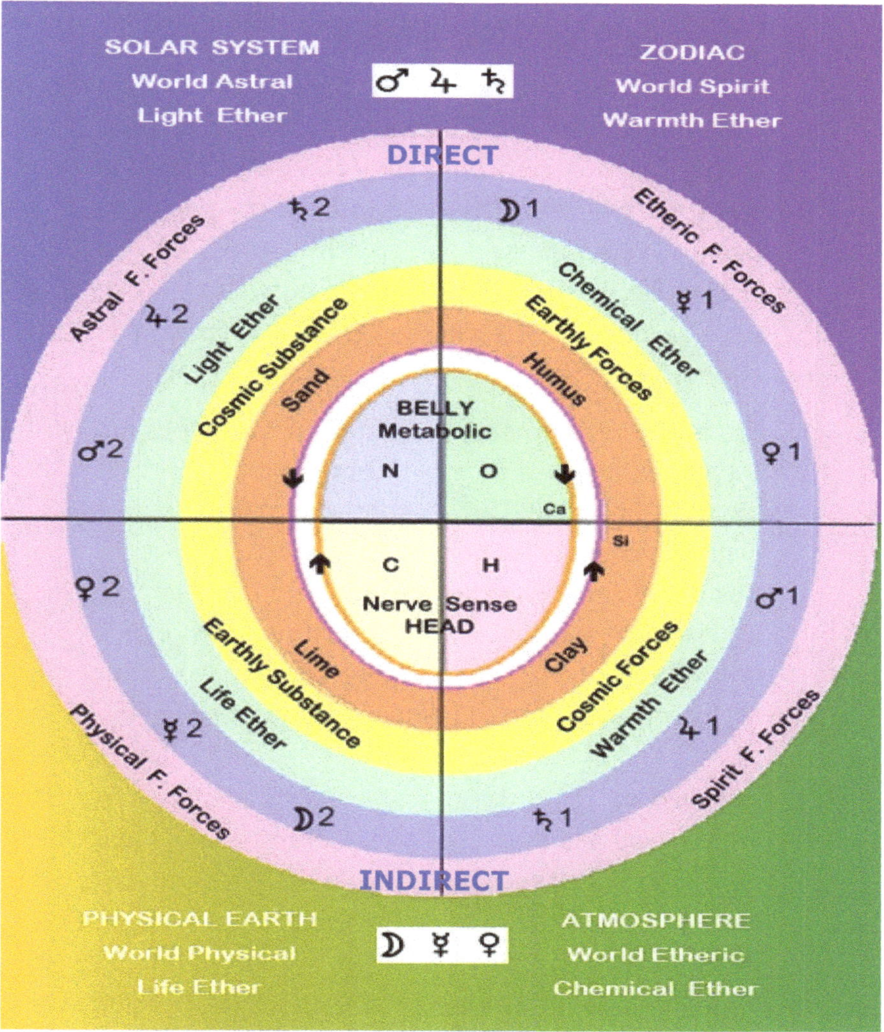

Lecture 3

In this lecture we are told how the 4 fold story of the Primary Energetic Activities works through the 5 'sisters of protein'. This provides **THE link of how the cosmic energies work into matter.**

This lecture however is incomplete and therefore somewhat confusing. The incarnating path of the Spirit via Hydrogen is not talked of, so it is easy, given some strong comments in the lecture, to confuse 'the carrier of the Spirit' as either Carbon or

Sulphur. Hydrogen is the 'first ' element, that drives the Stars, which are the source of Spirit force, and from which all other elements are derived. The medical lectures have this emphasis of Hydrogen as the carrier of the Spirit.

 Carbon is the physical carrier of ALL the energetic activities, as it provides the physical body, for them all to come into manifestation. While Sulphur is the 'oil' or facilitator that allows all of the energetic activities to work together into matter. Once these three understandings are clarified, this lecture becomes much more straight forward.

One of the highlights of the lecture is where RS tells of how Nitrogen draws Oxygen to Carbon.

"throughout Nature the oxygen bearing the etheric life must find the way to the carbon bearing the spiritual principle . How does it do this? What here acts as the mediator?

The mediator is nitrogen. Nitrogen (**Astral** *) directs the life (* **Etheric** *) into the form (* **Physical** *) which is embodied into the carbon. Wherever nitrogen occurs its function is to mediate between life (***oxygen***) and the spiritual element (***hydrogen***) which has first been incorporated in (with) the carbon substance. It supplies the bridge between oxygen and carbon - whether it be the animal and vegetable kingdoms, or in the soil."*

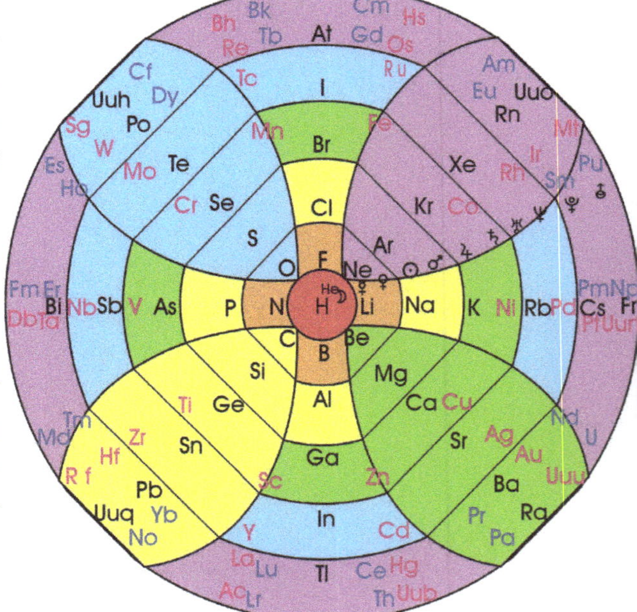

Translated this means that the Astral picks up Oxygen and gives it movement, while binding it to the Physical. This essential process is not clearly stated elsewhere in the course. However it is significant for lecture 4 and in understanding Equisetum. One observation of this relationship is in the Periodic table, where we see the order is Carbon, Nitrogen , Oxygen (orange ring)

Lecture 4

Two questions that arises in lecture 4 are 'how does the cowhorn work and what is occurring with the Astral comments'. We can note that within the nerve sense system we have the Spirit and Astral working. In the conversations of the cosmic and earthly nutrition streams in lecture 8, we are told, that depending on the strength of the Cosmic Forces / Spirit in the nerve sense region, we can determine how strongly the Astral related Cosmic Substance stream comes in through the senses, working down into the digestion. This initially is a World Astral activity that is transformed by the senses into a Internal Astral activity. Due to this nutrition image some people read the follow passage to mean that the World Astral comes from the front, through the horns and into relationship with the Etheric activity of the metabolism. Mirroring the 'front Sun to Moon rear' polarity talked of in the second lecture. They say horns are cosmic antenna, and hence the wonders of Horn Manure is all those World Astral Forces coming through the horns. This view has long been disputed, but nevertheless it persists, most recently by Dennis Klocek in 'Sacred Agriculture'.

Once we use the 4 fold ordering, we have a clear indication that the metabolism is sustained by the interaction of the Internal Etheric / Earthly Forces and Internal Astral / Cosmic Substance working together. Yes the Astrality has migrated from the nerve sense system, however this was a long time ago during the cows evolution. Now the Internal Etheric and Internal Astral have to work together for the metabolic system to work properly. It is this joint activity that moves forwards to the cowhorn, and is then reflected backwards. My studies into RS suggestions of Equisetum and kidney function, (10) show that it is an excess of World Astral activity in the metabolism that displaces the Internal Astral, which in turn stops activating the Internal Etheric, and the kidney stop working and shrink. The solution is to use the internal astral stimulating effects of Equisetum to push out the World Astral and begin pumping the Internal Etheric once more.

This quote is from the 1938 version of the course, as it is most easily accessed and out of copyright. The words in normal type are my 'clarifications', remembering that we need to read this description as an expression of the lecture 8 diagrams, of the internal organisation of the animal. RS manner of talking about the role of the Astral and Etheric within the digestive processes, is very common in the medical lectures, and forms the basis of several of his descriptions of illness.

"Have you ever wondered why it is that cows have horns, while certain other animals have antlers? It is a very important question. Yet what science has to say about it is quite one-sided and based on externals. Let us consider why cows have horns. I said that the forces within a living organism need not always be directed outwards, but can also be directed inwards. Now imagine an organic entity possessing these two sets of forces, but which is unformed and lumpish in build. The result would be an irregular, ungainly being. We should have curious looking cows if this were the case. They would all be lumpish and unformed, with rudimentary limbs as at an early embryonic stage. But this is not how a cow is constructed. A cow has horns and hoofs. Now what happens at the points where horns and hoofs grow? At these points an area is formed from which the organic formative forces, moving outwards from the metabolism, are reflected inwards in a particularly powerful way. ***There is no communication with the outside as in the case of the skin or hair; the horny substance blocks the way for these forces to the outside****. This is why the growth of horns and claws has such a bearing upon the whole form of the animal.*

Things are quite different in the case of antlers. Here the streams of forces, coming from the metabolism are not led back into the organism, but certain of them are guided for a short distance out of the organism; there must be valves, as it were, through which the streams localised in the antlers (we can speak of streams of 'force', just as we can speak of streams of air or liquid) can be discharged. A stag is beautiful because it stands in intense communication with its environment by reason of its sending outwards streams of metabolic forces; by this it lives within its environment and takes up from it everything which works organically in its nerves and senses. Hence the nervous nature of the stag. In a certain respect all animals which have antlers are suffused with a gentle nervousness. This is clearly to be seen in their eyes.

The cow has horns in order to reflect inwards the astral inspired Cosmic Substance and *etheric* inspired Earthly Force*, formative forces, which then penetrate right into the metabolic system, so that a activity in the digestive organism arises by reason of this radiation from horns and hoofs. In the horn, therefore, we have something which by its inherent nature is fitted to reflect* back the living Etheric and Astral streams into the inner life organs. The horn is something which radiates etheric life and even the astral element. Indeed, if you were able to enter into the cows belly, you would smell the current of Etheric-Astral life which streams inwards from the horns: and the same thing is true of the hoofs.*

Now this gives us a hint as to the measures we may recommend for increasing the effectiveness of ordinary stable manure. What is ordinary stable manure really? It is foodstuff which the animal has taken in and which up to a certain point has been assimilated by its organism, thereby stirring into activity certain dynamic forces in the organism. Its main use has not been to increase the amount of substance in the organism, for after having had its effect, it is excreted. It has become permeated with Astral /Cosmic Substance and Etheric / Earthly Force elements, active in the metabolism.

The Astral / Cosmic Substance *element has filled it with nitrogen-bearing forces and the Etheric / Earthly Forces element with oxygen-bearing forces. The substance which emerges as dung is permeated with these forces. Imagine now: We take this substance and pass it into the soil in some form or other . Thus we add to the soil an etheric-astral element whose proper place is in the belly of the animal, where it produces forces of a plant-like nature. For the forces which we produce in our digestive tract are of a plant-like nature. We should be extremely thankful that we get such a residue as dung, for it carries etheric and astral forces from the interior of the organism out into the open. These forces remain with it, and it is for us to keep them there. In this way the dung will act in a life-giving and also astralising way on the soil, not only on the water element in it, but especially on the solid (earthly) element. It has the power to overcome what is inorganic in the earthly element. Now what is passed over to the soil will necessarily, of course, lose the form it originally had when taken in as food, for it has to go through an inner organic process in the metabolic system. There it enters upon a phase of decomposition and dissolution. But it is at its best just at the point where it begins to dissolve through the workings of its own astral and etheric elements. It is then that the parasites, the micro-organisms make their appearance. They find a good feeding-ground in which to develop. This is why the theory arose that these parasites are themselves responsible for the virtues in the manure. But they are only indications of the condition of the manure. If we think that by inoculating the manure with these bacteria we shall radically improve its quality, we are making a complete mistake. Externally there may seem at first to be an improvement, but in reality there is none. I shall deal with this point later. For the moment, let us continue with the matter in hand.*

Let us put manure just as it comes to hand into a cow-horn, pressing it full, and bury it at a certain depth - say I to 2 feet deep according to the soil which should not be too sandy or clayey. We can choose any spot where the soil is in good heart. Now by thus burying it with its filling of manure, we preserve in the horn that function which it would normally exercise in the cow's body, that is the reflecting and concentrating *of the internal life-giving and astral elements. Through the fact of its being surrounded with earth, all the currents of Etheric and Astral forces stream into its interior,* from the open end. *These forces attract all the Astral and Etheric elements from the surrounding soil, and the manure contained in the horn becomes inwardly quickened with these forces in the course of the winter season when the earth itself is most alive. For the earth is most inwardly alive during the winter. All these living forces are preserved in the manure and thus there is a highly concentrated, life-giving manuring force in the contents of the horn. Then (in spring) the horn can be dug up and its contents removed. Those of you who were present at Dornach when last we made this experiment will remember that you were able to convince yourselves of the fact that when the manure was removed it was completely odorless. It was quite striking. The manure no longer smelt at all, though naturally it began to do so a little when it was mixed with water. This shows that all its odour had been concentrated and worked up within it. You have here a tremendous astral and etheric power which you can utilise by taking the content of the cow-horn after its period of hibernation and diluting it with water which perhaps should be slightly warmed.*

Lecture 5

Here are the various planets relationships and the energetic activities of the preparations. I like Dr Lievegoed's planetary associations, which support and build upon the energetic activities of the three preparations that RS mentions in this lecture. I added the suggestions for 506 and 507.

Hugh Lovel has championed a slightly different organisation to this. He says it comes from Maria Thun. He says Saturn is Equisetum, Mars is Valerian and Sun is Stinging Nettle. I can see why they would consider that however my study into Equisetum, with its working into the metabolism and particularly the kidneys, places it more as a 'middle' Venus Mars preparation. It's primeval, pre flowering plant origins and stem

SPIRIT Indirect Outer Pl. 1 Warmth Cosmic Forces Clay	**Cambium** **Salamanders** ↓	**Saturn**	P	**Valerian**	**Strengthens the Spirit**
	Sylphs	**Jupiter**	S	**Dandelion**	**Merge Spirit and Astral towards the Physical**
ASTRAL Direct Outer Pl. 2 Light Cosmic Substance Sand	**Life Sap**	**Mars**	Cl	**Nettle**	**Harmonise the Astrality**
		Sun	Ar	**Equisetum**	**Astral stimulates the Etheric within the metabolism**
ETHERIC Direct Inner Pl. 1 Chemical Earthly Forces Humus	**Undines** ↓	**Venus**	Na	**Yarrow**	**Opens the Etheric to the Astral**
	Gnomes **Wood Sap**	**Mercury**	Mg	**Chamomile**	**Stimulates the Etheric**
PHYSICAL Indirect Inner Pl. 2 Life Earthly Substance Cations	↑ **500**	**Moon**	Al	**Oak Bark**	**Etheric binds to the Physical**
		Earth	Si	**Quartz**	**Spirit binds to the Physical**

dominance takes it away, from the seeding processes associated with Saturn. With Stinging Nettle we can see from RS words and its action as a remedy that it works particularly to calm the Astrality. Anyway their suggestion are worth exploring for yourself.

Lecture 6

In this lecture we are given insights into Fungal and Pest problems. The passage where RS describes fungal rot problems has some very confusing indications. Again looking at this passage through this fourfold filter offers some intriguing suggestions.

"There remains for us one more subject to consider: the so-called plant diseases. Actually this is not the right word to use. The abnormal processes in plants to which it refers are not "diseases" in the same sense as are those illnesses which afflict animals. When we come later on to discuss the animal kingdom, we shall see this difference more clearly. Above all, there are not processes such as take place in a sick human being. For actual disease is not possible without the presence of an astral body. In man and animals, the astral body is connected with the physical body through the etheric body and a certain connection is the normal state. Sometimes, however, the connection between the astral body and the physical body (or one of the physical organs) is closer than would normally be the case; so if the etheric body does not form a proper "cushion" between them, the astral intrudes itself too strongly into the physical body. It is from this that most diseases arise.

Now the plant does not actually possess an astral body of its own. It does not therefore suffer from the specific forms of disease that occur in men and animals. This is the first point. The next point is to ascertain what actually causes the plant to be diseased. Now, from everything I have said on this subject, you will have gathered that the soil immediately surrounding a plant has a definite life of its own. These life forces, in the form of Earthly Substance *are there and with them all kinds of forces of growth and tender forces of propagation not strong enough to produce the plant form itself, but still waiting with a certain intensity; and in addition all the forces working in the soil under the influence of the Moon and mediated through water. Thus certain important connections emerge, in the first place you have the earth, the earth saturated with*

water. Then you have the moon. The moon beams, as they stream into the earth, awaken it to a certain degree of life, they arouse "waves" and weavings in the earth's etheric element. The moon can do this more easily when the

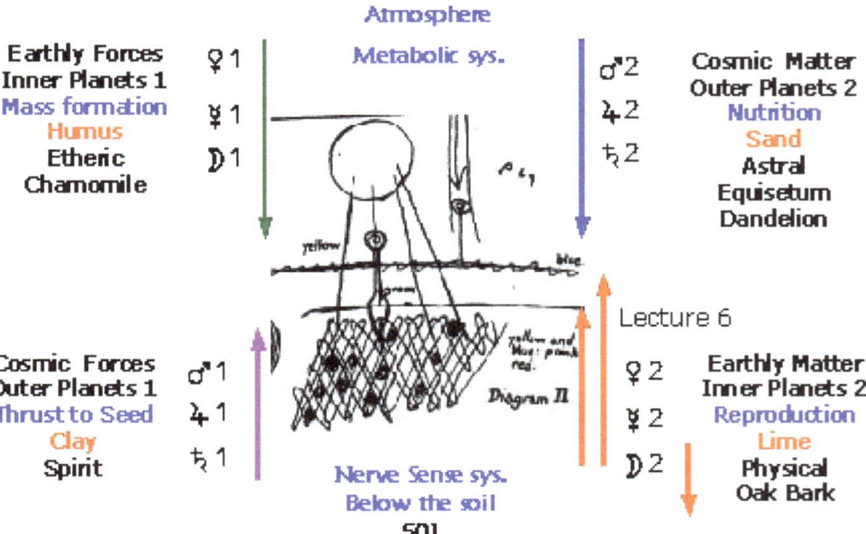

earth is permeated with water, less easily when the earth is dry. Thus the water acts only as a mediator. What has to be quickened is the Earth itself, the solid mineral element. Water, too, is something mineral. There is no sharp boundary, of course. In any case, we must have lunar influences at work in the earth, via the Earthly Substance. *Now these lunar influences* and the Earthly Substance *can become too strong. Indeed this may happen in a very simple manner. Consider what happens, when a very wet spring follows upon a very wet winter. The lunar force enters too strongly into the earth, which thus becomes too much alive. I will indicate this by red dot's. (See Diagram No. 11). Thus if the red dots were not here, i.e. if the earth were not too strongly vitalised by the moon, the plants growing upon it would follow the normal development from seed to 'fruit; there would be just the right amount of lunar force distributed in the earth to work upwards*, upon the Cosmic Force stream, *and produce the requisite fruit and seed. But let us suppose that the lunar* Earthly Substance *influence is too strong - that the earth is too powerfully vitalised - then the* Earthly Substance *forces working upwards become too strong, and what should happen in the seed formation, occurs earlier. Through their very intensity the Earthly Substance forces do not proceed far enough to reach the higher parts of the plant, but become active earlier and at a lower level.* This is caused by the Earthly Forces acting alone, without the Cosmic Forces, and indeed being so strong as to suppress the upward Cosmic Forces stream. *The lunar influence has the result that there is not sufficient strength* within the Silica stream, to meet the Cosmic Substance active above, *for seed formation. The seed receives a certain portion of the decaying life, and this decaying life forms another level above the soil level. This new level is not soil, but the same influences are at work there. The result is that the seed of the plant, the upper part of the plant becomes a kind of soil for other organisms; parasites and fungaloid formations appear in it. It is in this way that blights and similar ills make their appearance in the plant. It is through a too strong working of the moon* and the Earthly Substance *that forces working upward from the earth are prevented from reaching their proper height. The* siliceous powers of fertilisation and fructification depend entirely upon a normal amount of lunar influence. *It is a curious fact that abnormal developments should be caused not by a weakening but by an increase of lunar forces. Speculation might well lead to the opposite conclusion. Looking at it in the right way shows that the matter is as I have presented it. What, then, have we to do? We have to relieve the earth of the excess of lunar forces in it. It is possible to relieve the earth in this way. We shall have to discover something which will rob the water of its power as a mediator and restore to the earth more of its earthiness, so that it does not take up an excess of lunar forces from the water. This is done by making fairly concentrated brew (or tea) of equisetum arvense (horse-tail), diluting it and using it as a liquid manure on the fields for the purpose of fighting blight and similar plant diseases. Here again only small quantities are required; a homeopathic dose is generally sufficient. As you will have realised, this is precisely where one sees how one department of life affects another. If, without indulging in undue speculation, we realise the noteworthy effects produced by equisetum arvense upon the human organism by affecting the function of the kidneys, we shall have, as it were, a standard by which to estimate what this plant can achieve when it has been transformed into liquid manure, and we shall realise how extensive its effects may be when even quite a small quantity is sprinkled about without the help of any special instrument. We shall realise that equisetum is a first-rate remedy. Not literally a remedy, since plants cannot really be ill. It is not so much a healing process as a process exactly opposite to that described above."*

The suggestion for Equisetum suggests RS is strengthening the Internal Astral or its surrogate the Cosmic Substance in the metabolic region, to push down and both suppress the overly active Earthly Substance, but also to encourage the sleepy Cosmic Forces from below. However if we reflect upon lecture 2, RS gave very clear instruction that we could strengthen the Cosmic Forces through the use of clay. We also need to reflect on the comments RS made about the functioning of the Oak Bark prep, in that it helps to pull and overly active Earthly Substance / Moon activity back into the physical body, without causing any shocks.

Hence a combination of Clay, Equisetum and Oak Bark becomes a suggestion against rot fungus. My experiments have shown this to be a effective remedy.

This use of Equisetum appears to be the opposite remedy to RS Bidor preparations, made from Quartz, Iron and Sulphur. This aims to work from the nerve sense system / head and push back an overly active metabolic process, which is working too strongly upwards and causing headaches. See (10) for a more detailed investigation of this matter.

Pest Control

Also in lecture 6 we find these comments.

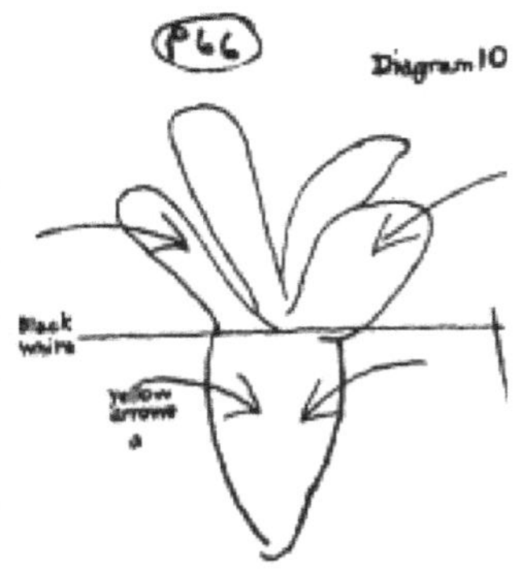

"Now I am going to tread on very thin ice and take an example very near home. I am going to talk about the nematode of the beetroot. The outer signs of this disease are a swelling of root fibers and limpness of the leaves in the morning. Now we must clearly realise the following facts: The leaves, the middle part of the plant which undergo these changes, absorb cosmic influences that come from the surrounding air, whereas the roots absorb the forces which have entered into the earth and are reflected upwards into the plant. What, then, takes place when the nematode occurs? It is this: The process of absorption which should actually reside in the region of the leaves has been pressed downwards and embraces the roots.

Thus if this (Diagram No. 10) represents the earth level, and this the plant, then in the plant infested with the nematode the forces which should be active above the horizontal line are actually at work below it. What happens is that certain cosmic forces slide down to a deeper level; hence the change in the external appearance of the plant. But this also makes it possible for the parasite to obtain under the soil (which is its proper habitat) those cosmic forces which it must have to sustain It (the nematode is a wire-like worm). Otherwise it would be forced to seek for these forces in the region of the leaves; this, however, it cannot do as the soil is its proper environment. Some, indeed all, living beings can only live within certain limits of existence. Just try to live in an atmosphere 70 degrees above or 70 degrees below zero and you will see what will happen. You are constituted to live in a certain temperature, neither above nor below it. The nematode is in the same position. It cannot live without earth and without the presence of certain cosmic forces brought down into it. Without these two conditions it would die out.

Every living being is subject to quite definite conditions. And for the particular beings with which we are dealing, it is important that cosmic forces should enter the earth, forces which would ordinarily display themselves only in the atmosphere around the earth. Actually the workings of these forces have a four-year rhythm. Now in the case of the nematode, we have something very abnormal. If one enquires into these forces, one finds that they are the same as those at work on the cockchafer grubs; and as those, too, which bestow on the earth the faculty of bringing the seed potato to development. Cockchafer grubs as well as seed potatoes are bred by the same forces, and these forces recur every four years This four yearly cycle is what must be taken into account not with regard to the nematode but with regard to the steps we take to combat it."

In this case we have an overly active World Astral / Cosmic substance activity causing the pest problem, which he really did not provide the solution for. He went on to talk of peppering. However this does not provide a solution to this energetic problem. Peppering does not alter the environmental energetic activities. It cuts off a specific reproductive stream to a specific part of a specific species. Experience shows , you might rid yourself of one pest via peppering, but another of the same natural niche will replace it. His suggestion for altering the environment to fend off pests is achieved by other means, such as specific BD preps and specific chemical elements. Nevertheless he has given us this very important image about the energetic reason for some pest attacks. It would seem appropriate to pull out the overly active Astrality with Stinging Nettle and stimulate the Etheric inspired Earthly Substance activity within the soil, with Horn Manure, Chamomile and Yarrow.

These are the most contentious and influential passages in the course, however if there are specific passages you do not understand, I suspect the fourfold view will help. Have a look about what I say about it in my commentary, online. Or let me know and I will see if I can shed some light on it

Lecture 8

This lecture speaks for itself and gives a very clear application of the four energies working within animals. Given the fourfold basis of the whole course presented in this essay, lecture 8 is an essential and completing part of the story.

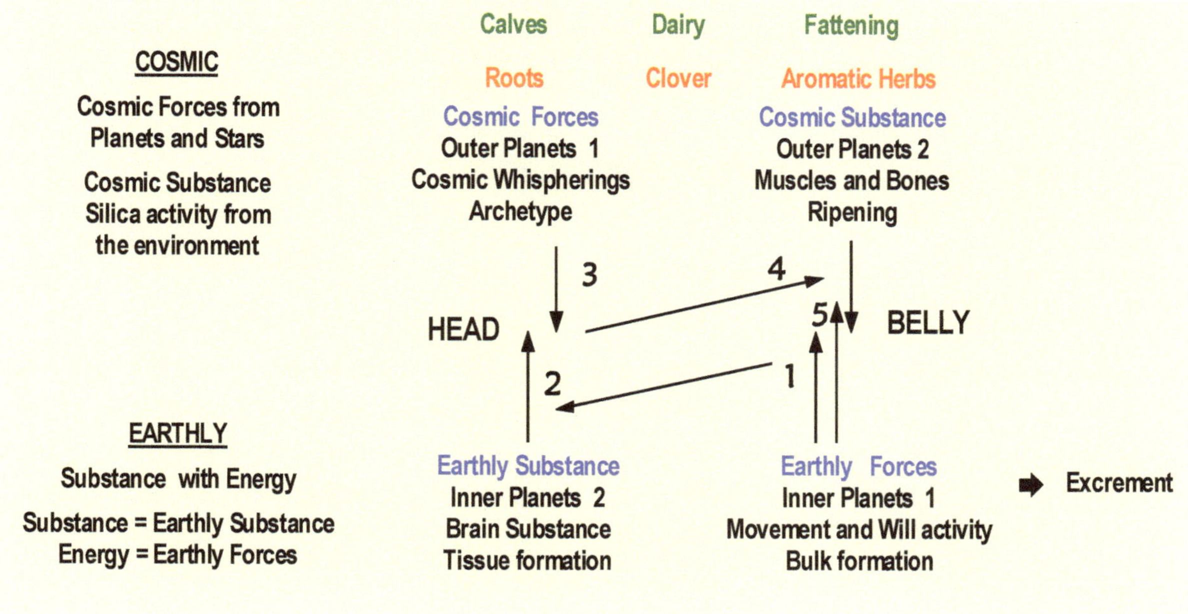

Appendix 1

An Astrological View of
Dr Steiner's Agriculture Course

As a long time student of this course, I have been amazed at the levels of confusion it has caused, over the last 100 years. Yes it is a very dense course, and as my re edit showed, it is quite 'mixed up'. The two basic stories — the Energetic Activities and the Physical Formative Forces (PFF) — are often intertwined throughout the text, so that it is very difficult to know which story is being told, when. Indeed the PFF story is so obscure it has been completely lost to many people. I hope I have shown that if we use the Medical lectures and their agreed 'framework', we are able to find the structure RS uses for his agricultural indications. Interestingly, it is NOT the structure you will find most BD folk using. Instead of, the 4 energetic activities taking place within a 3 fold physical organism— outlined in Lecture 2 - we have most folk only making it to a 3 fold image. An expansive pole , a contractive pole and a middle. In this 'cut down' form lots of things do not make sense, and these confusions are therefore 'left behind'. With this confusion about the course, and a lack of a consensus authorative understanding, people are free to make up whatever stories they want. Given the isolation of farmers from each other, over time we have several 'islands of delusion' floating about the planet, pushing their 'local gurus' ideas. The internet is allowing us to see all these different views of BD, however the continued lack of an authorative understanding of RS message, at its centre, means no serious critiques of these 'islands' can be undertaken. I have begun this critique process, from my isolated position on the periphery of the planet, however it is like speaking into a vacuum. No one need listen and Yes, things do take time to comprehend and seep into the collective……..

How have we gotten here, when other RS initiatives have managed to find their 'authorative understanding' around which to coalesce?

My interest in Astrology has been very useful in finding RS underlying order, and moving it along some, however I had not gotten around to asking the question of **what is contained in the astrological moment at the beginning of the Agriculture Course**. In Astrology we place importance upon the 'moment of beginning', be it the birth of a person or the initiation of a project. In this moment we have the seeds of what will unfold in that person's or project's 'personality and life'. So the moment of the start of the first lecture of the Agriculture Course, should contain information about the 'Life of the Course'. This will be what is provided, how it is received and how it will unfold through time.

So the other day I did ask this question………and there in front of me is the story I have told you above.

I believe I have the right starting time, from Peter Selg's book on the course (ISBN978 1 906999 08 7) as **11 am on the 7th June 1924 at Wroclaw, Poland.** The chart of that time is on the next page. There is a lot going on in there, and so it helps to simplify this by looking at the 'conversations' going on between the planets. I like to

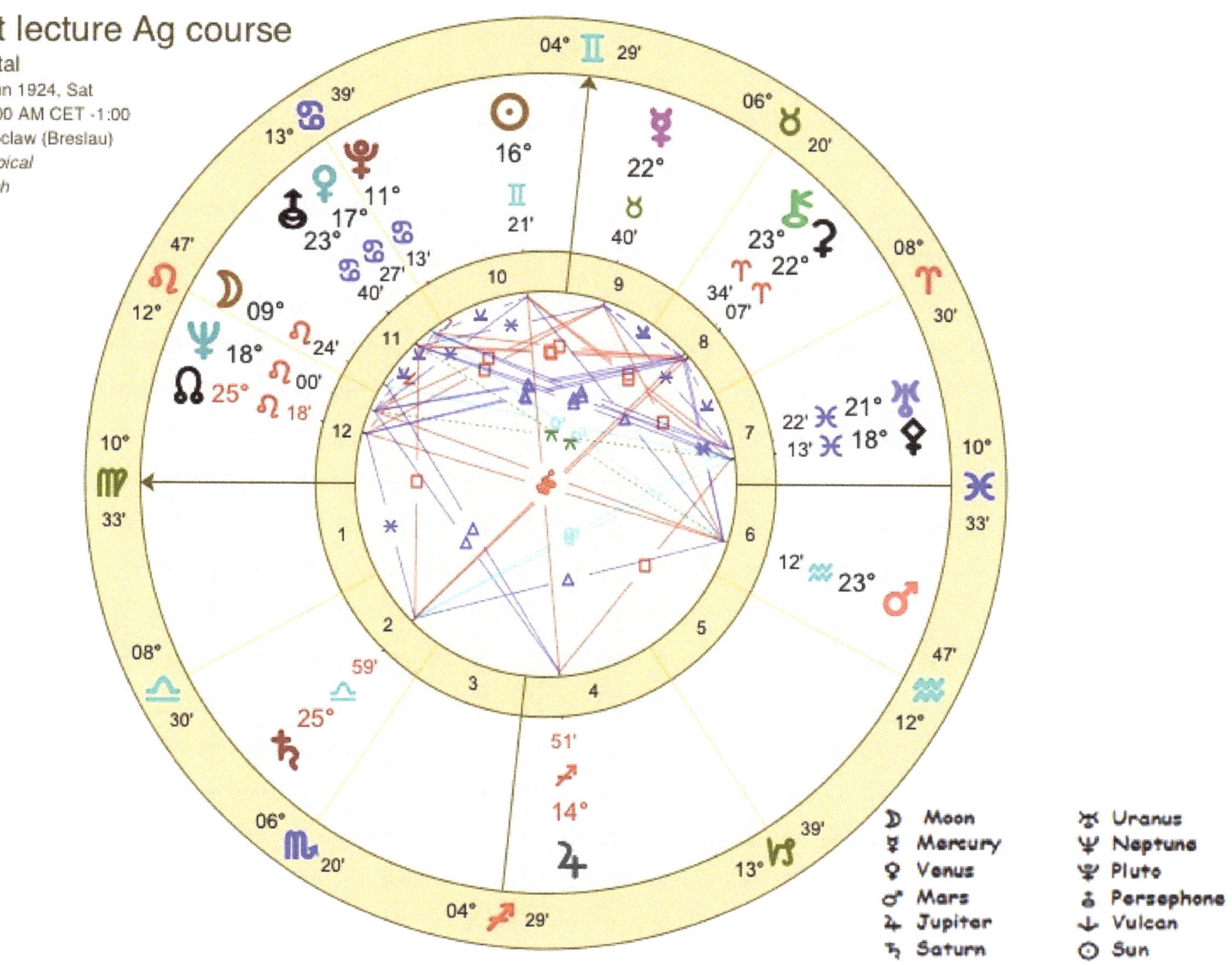

look at charts as if they were an executive team in a business. Each planet is a head of a department, and this picture is them sitting around the conference table, talking to each other. The **lines are the Types** of conversations they are having with each other. The **Planets are What** they are talking about. Through the process of pulling this image apart, into its structural groups, we can listen to all the different conversations going on here. One way to start is through the different types of conversations. Like in music, with its different tones, so we have harmonious and discordant / non harmonic conversations. In the charts, the **Blue lines are the harmonious 3rd harmonic**, while the **Red are the discordant, 4th harmonic** conversations. The **light blue are especially good 5th harmonic** conversations. These are 'the gifts' of the chart.

Sun, Moon, Ascendant and Midheaven

But first, it is 'traditional' to begin an analysis with a look at these four points. These provide an image of the conscious intention, the unconscious intention, the persona or how these are seen, and the impact upon the world.

The **Sun** in Gemini in the 10th house indicates a youthful, bright, even playful intention to achieve a certain goal, however the twins image suggests there will be two different approaches to this course. We can see today a desire to keep it simple and materialistic eg the preparations work as stimulators of the chemicals RS mentions with them. The second approach presents a spiritual energetic science that appears to still need some clarification and development.   The sabian symbol of the 17th degree of Gemini is **"The head of a robust youth turns into that of a mature thinker".** Is this the dual Gemini nature, and our challenge is to mature our thinking and perception beyond the 'easy' materialistic approach?

The **Moon** in Leo in the 11th house indicates an inner confidence with ones vision for the future of ones community. This unconscious urge shows the intention of the course for the good of humanity. The sabian symbol for 10 Leo is **" Early morning dew sparkles as sunlight floods the field"** . The interpretation of this is stated as " The exalted feeling that rises within the soul of the individual who has successfully passed through the long night which has tested his strength and his faith". The Moon being with Neptune shows the deeply spiritual and intuitive nature of the movement. It can also give both Sun streams an arrogance of belief in their own correctness, while Neptune adds a religious fervor to this belief.

The **Ascendant** at the 11th degree of Virgo (along with Saturn rising) suggests a very measured and precise manner of expressing one self. Somewhat reserved but also sure of ones details. This shows in the BD movement wishing to be seen as conservative, scientific and not scarey. The sabian symbol for this degree is " **In her baby a mother sees her deep longing for a son answered"**

The **Midheaven** is 5 degrees Gemini, shows the goal, higher aspirations and the impact in the world of the event. The Symbol is **"A revolutionary magazine asking for action".**

The North Node at 25 Leo is a "**A rainbow after a heavy storm."** The north node suggests the destiny of the course. A possibility of a new beginning after the onslaught of toxic agriculture.

The Planets Discussions

Given the human penchance for 'nice things' first, I will start with the positive indications of the course. The chart on the next page has only the 3rd harmonic (positive) relationships between the planets. I have also added three other celestial bodies. The asteroids Ceres (agriculture) and Pallas Athene (leadership) and the hypothetical planet Persephone (Transpluto) which I have found very accurately predicts how something works for the whole of humanity, and hence its international influence.

The next step is to see who is talking to who, and wherever there is a pattern of interactions, there is a complex conversation going on. However we can consider this just one 'flowing' conversation. There are various emphasis points within the conversation, but it is still one conversation.

Looking at this chart, we can see that all of the blue lines, more or less connect up with each other. I can see two main conversations in this picture, who are connected to each other. So they can all hear each other.

Blue 1—The big rectangle

Saturn, Mars, Ceres, Chiron, North Node, Neptune, Sun, Venus, Persephone

Blue 2 - Small triangle on the right

Uranus, Pallas Athene, Chiron, Ceres, Mercury, Venus, Persephone

Both of these have interconnecting links, however it is better to see the two parts, first.

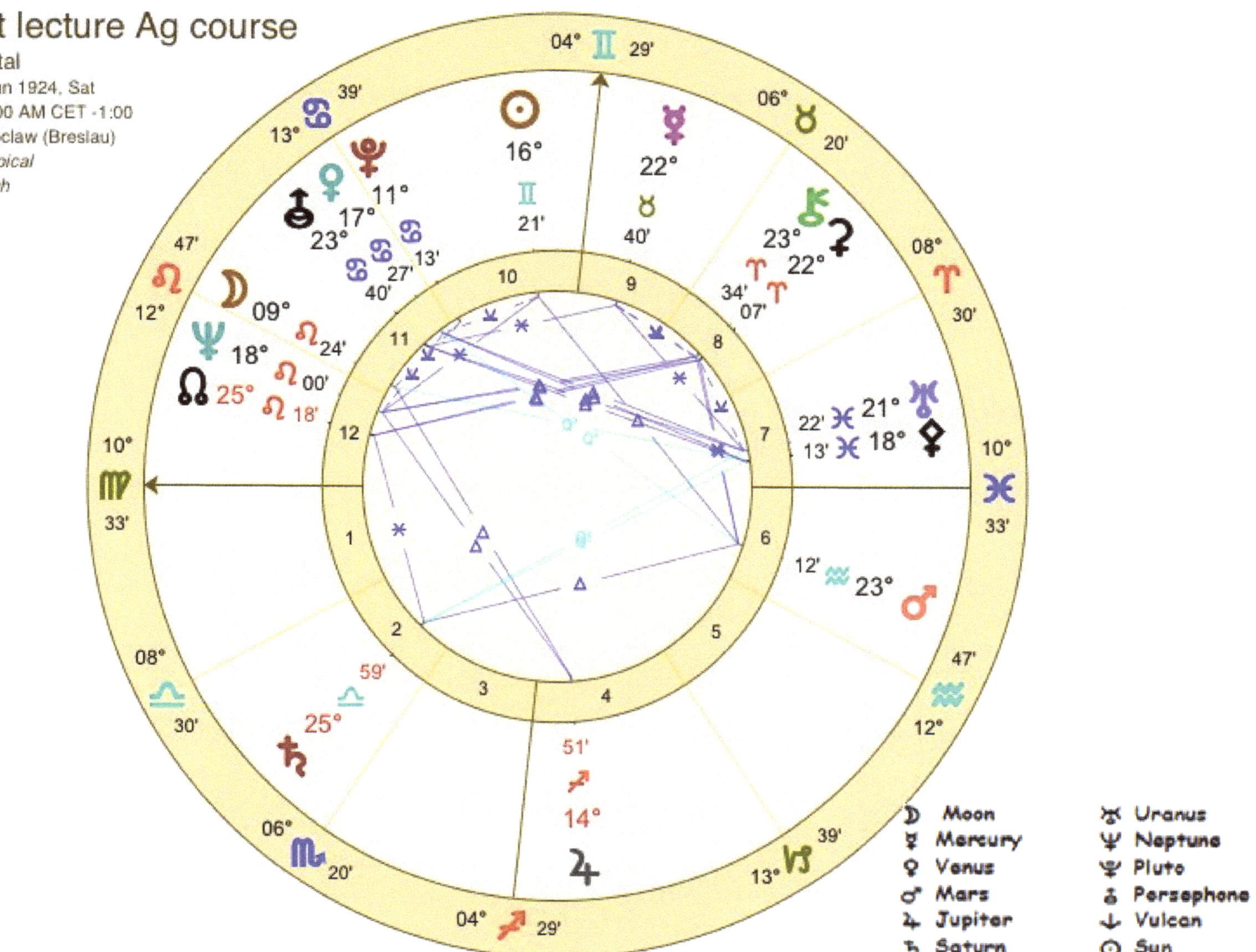

Blue 1

These pieces of the conversation show their positive sides. Given this is a agriculture conference we can start with **Chiron and Ceres in Aries**. Chiron is the healer, while Ceres is the goddess of agriculture and in Aries, the first sign, so we hear them talking of **highly innovative ways to heal global agriculture.** Following the big rectangle clockwise, **Mars and Saturn** both are male 'lets make this practical' sort of guys, and want these ideas to be acted upon. The air sign placements of these planets, do indicate the ideas are 'lofty' , and will take a little effort to make practical. The fourth leg of this rectangle is **Neptune and Node**.   Neptune is the idealist in this conversation, and while he likes the practical nature of Mars and Saturn, he is bringing in the spiritual and humanitarian points of view to the table. The Node in this mix is saying **'the time has arrived'** for this practical vision to be presented. When we focus on Neptune, we see it has a positive side conversation going with the Sun. The Sun is the conscious individual, which suggests that the message of this conference can be consciously grasped and put into practice. While here we can mention the Sun has a 'small' conversation going with Venus Persephone and Neptune. This talks of good will amongst the international participants within the movement, as can be seen every year at the Dornach conferences.

Blue 2

The **second pattern** sitting alongside this first one, is based on **Mercury**—the communicator, **Uranus**— the innovative thinker, **Pallas**—leader, and **Venus Persephone** International people and **Ceres Chiron** - Healing agriculture; All these indicate the innovative , multi layered and revolutionary nature of the ideas that are presented. It also brings the 'open to all people' nature of the knowledge, and the cosmopolitan community we are.

Jupiter is standing on a leg of its own and in Sagittarius, so it is offering many stimulants to the 'lofty' idealistic side of the conversation. It adds to the humanitarian socially conscious ideas, but as we will see shortly, it is a little over enthusiastic.

Light Blue 1
The **third pattern** we can see here is the light blue lines between **Saturn, Uranus Pallas and the Moon**. So far the Moon has not featured. This structure, indicates a very special practical inspiration which is carried somewhat intuitively. This is the wonders of Goethean perception, but it can also suggest **the women of our movement will play a significant role in its innovative development and leadership**. We do not have to look far to see this. This is one of our strengths………

With the four images, of the Sun, Moon, Ascendant and Midheaven, sitting at the base of the 'positives' outlined above, I see the fulfillment of RS journey. He has come of age as a striving human, and presented a gift of renewal for Agriculture, as an expression of his skill and competence, as a seer of truth, meeting a need of the time.

Bill Watson offers— *I sometimes check out the Part of Fortune, (Ascendant + Moon — Sun) which also has an apt symbol for this event:* **"A youth carries a lighted candle in a devotional ritual".** *–The educative power of ceremonies (and symbols) which impress the great images of a culture upon its gathered participants." Consider that this lecture was delivered 95 years ago, and still has educative power through its symbolic meanings and stirring rituals.*

Bill offered 4 other astrological references that supported the basic themes outlined in my assessment.

So all very good………..

But…………..and there is always a but…….**there be demons lying under that log**……. which brings us to the **tensions in this chart**. The chart above, highlights the tensions. In it we can see three fairly defined groups of planets. While they do not talk to each other, they do converse through the positive relationships. This suggests resolve is possible, into a cohesive 'personality'. Nevertheless 'the dragons' need to be taken seriously and bought to consciousness , if they are to be turned into positives.

Red 1 — Neptune, Moon, Mercury, Mars (Pluto Node)

Red 2 - Sun, Uranus, Pallas Athene, Jupiter

Red 3 — Saturn, Persephone Venus Ceres Chiron.

Red 1 — The Curse
Mercury is communication and ones personal ideas, **Mars** is the assertiveness one puts behind those ideas, while the **Moon and Neptune**, indicates those ideas are hard to bring to awareness, and can be highly imaginative, and easily heading towards complete delusion. The **Pluto** tension to Node off this, suggests this tendency can have

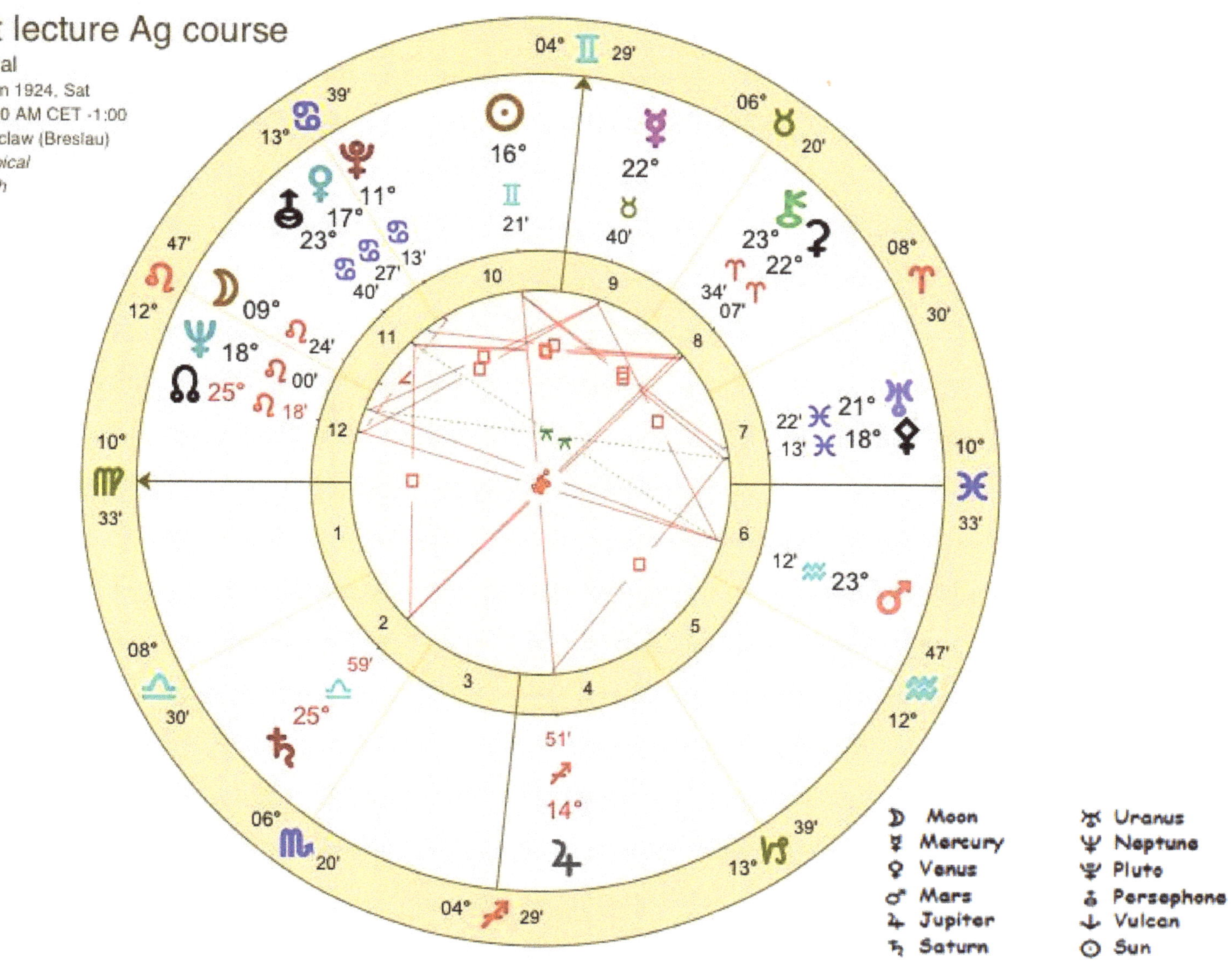

a debilitating effort upon the ultimate destiny of the endeavor.

This aspect suggests (a) people will have great difficulty understanding what is being said, as they perceive it to be too vague, abstract, waffly or unattainable. (b) people will take the information provided in the course, and make up all manner of delusory stories from it. With all of this encompassing the Moon in Leo, it suggests peoples confidence in their own inner vision of what is said, will take precedence over what might be consciously presented to them.

Add to this the basic tenet of RS and Goethean perception, that we must honor the individuals perception, and we have 'free play'. Anything is possible. This aspect plays into RS own 'unaspected' Mercury Neptune in Pisces. He is already often, fuzzy and difficult to follow, so adding this 'natural' tendency to the course chart, creates a very strong indication of 'a problem'.   **Herein lies the largest danger** of this time for starting the course. (Interestingly, as I write this Mercury is conjunct Neptune in Pisces. Is this a phenomena whose time has come for it to be bought to consciousness? Also I recently finished my Enzo Nastati critique, where I discussed the effects of his Mercury Neptune conjunction.)

A credible science must have some central pillar it can 'bounce off'. My suggestion is RS medical lectures and the worldview the very substantial Steiner medical movement uses, is that pillar. Some will say Anthroposophy should be that pillar, however I suggest this larger body of work, with its lofty unprovable stories, feeds into this aspect's delusory potentials. The medical lectures are focused upon practical applications, and so has boundaries and patterns of thought, to guide our observation.

Red 2

Plays into this tendency. **Sun, Jupiter, Uranus, Pallas** are all strong 'I' statements. These planets provide a very strong self confidence in ones ability to understand multiple levels of understanding and applications. This is a person who can boldly take leadership towards their implementation, without concern for what anyone else might say. The multi dimensional nature of the information and its implications, can seem 'mind boggling' to the 'average' person. 'Its all too far out'…..yet to people that can cope with this 'big brain' aspect, they are provided with a work of genius, that can be applied to many fields of application.

The positive flip side of this, is the ability to carry on with ones ideas, regardless of the outside worlds persecution, due to seeing 'a bigger picture'. Which we might say is a very necessary 'arrogance' one needs to birth a reformative movement. However if this arrogance is used to support a delusion, and we have someone, who is not listening to critique, then we have a problem.

Disclosure: I have this planetary combination in my birth chart, and so my efforts are a good example of this aspects various manifestations, and probably why I can chase down the wild ideas this aspect develops.

Red 3 - holds its own dragons.

The key players in this conversation are **Saturn and Venus**. While these might be considered in wide aspect, and therefore a weak influence, bringing in Persephone, Ceres, Chiron and by association Pluto, we do get a significant 'beast'.

Saturn and Venus in tension leads to an experience of **emotional abandonment and isolation.** In a person, they feel 'unloved' and 'unappreciated'. Add Pluto and we have feelings of persecution, and with Ceres, these come from within the family. While Persephone says it is an international family. Hence members of the family easily feel abandoned and persecuted.

Again we do not have to look far for this phenomena. Herein lies the many hurt feelings between all the various sects and individuals, obvious within the international movement. Place this within the setting of individuals and sects, being arrogant about their delusions, and we have a right royal festering sore.

We can see this in the immediate separations that occurred after RS death, in the general Anthroposophical society, and flowed into the BD movement, which appears to have given Wachsmuth free reign to lead BD down the 'distracting' Ethers road. To the regional 'divisions' we saw in Australia (Podalinsky and the rest) , New Zealand (Proctor and Atkinson) , and in Europe (Erbe, Fenstalin, and Nastati and the rest) to name a few. I do not know of the details in the USA, but I do know of several ongoing 'hurt feelings', over there as well.

At the seat of all these disagreements often sits **'the arrogant holding to individual**

positions', without any central truth to help with finding a resolution . The fairly common 'rule by isolation' technique of the central groups, feeds this demon and does little to consciously address this.   However when deluded, arrogant people 'feel isolated', they go off and form their own 'BD kingdoms'. Which leaves the people left in the middle completely confused, and, with many different stories that do not mesh together. With no means of integrating these wildly divergent views, we come to 'lets keep it simple', add some *spiritus fermenti*, and we have the naive BD 101, we see running about today.

This activity can be used for good, even though it is a tension. With Ceres - the goddess of Agriculture - involved it sways the conversation to indicate **an international group of long term committed friends working to heal agriculture,** even if it is a thankless task.

Where to

A few years ago this inability for a central story to act as pillar of truth, lead to my article "Any Old Story Will Do".

Herein lies Biodynamics 'curse', if we allow these unconscious nesses to carry on, why would anyone take us seriously. Everyone needs to **Come and Scratch their story upon, the Medical Lectures Pillar..........**please believe RS medical lectures. They are very practical, there is over 100 of them, and they have been worked upon by very conservative professional doctors, for 90 years. They know what they are talking about and what RS said. I am confident my story fits within their framework, which is why I present my efforts, and critique others, where I see glaring errors.

Yes, we have an astonishingly good gift for the renewal of Agriculture, but we need to find a compassionate way, to achieve a consensus reality, upon which we can move forward. **It is not OK that 'Any Old Story Will Do', we owe RS much much more than that...........**

Critique

Given the challenging, and potentially subjective nature of Astrological interpretation, I have given this essay to a competent astrologer to critique for its 'accuracy'. His assessment say -

*As a professional astrologer and bioenergetic medical practitioner, with forty years' experience, I found Glen's reading of the event chart for the first of Dr. Steiner's lectures on Agriculture, **to be both accurate and fascinating**. As I explored the chart it occurred to me that the date and time of that first lecture may have been "elected", i.e. specifically chosen for the quality of energy, rich with meaning. I believe the chart provides deep insights into the way forward for BioDynamic Agriculture, as expressed in Glen's concluding remarks on the subject.*
Prof. W.D. (Bill) Watson
Auckland, New Zealand

Regarding Bill's comment of the first lecture being a conscious election, we can note that the first lecture was 3 days prior to the second lecture, and the rest of the course. Dr Steiner's astrologer, Dr Elizabeth Vreede, was in attendance at these lectures.

The Signs and the Constellations?

'The Directors Cut - 6.9.2020'

The biodynamic community has long had a general difficulty discerning the difference between the Tropical 'Sign' zodiac and the Sidereal 'Constellation' zodiac. In the 1980s I wrote a book on this subject, 'Spiral Astrology', which can be had for free on my website. (6) It appears the same old 'confusions' about this topic that were around in the 80s, are still within our community. The problem appears to be we have two things that on the surface, look the same and use the same words, however they are very different in both their 'astronomical reality' and therefore their context of use. Given the Biodynamic community lacks competent Astrologers amongst their number, neophytes are usually left to make up somewhat odd stories.

Like everything in Biodynamics, we can undergo a 'Goethean observation' of the topic, to identify some truths about it. So first we must observe 'the issues'.

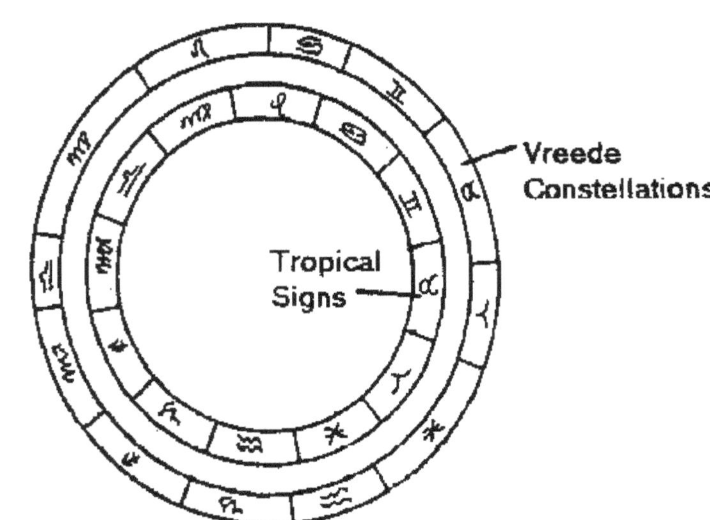

The Issue is, we have the 12 constellations in the sky, in front of which the Sun moves. These are based upon real stars. Then 2000 years ago, Ptolemy started talking about a very specific division of the Suns path itself, using this 12 fold division of the constellations, as a reference. This division starts each year at the same specific point, the Spring equinox of the northern hemisphere. Over the years these two 'organisations', no longer line up. During this time the equinox event has moved along the constellations. The Signs therefore have broken free of the constellations at 1 degree every 72 years, so that now the spring equinox at the northern hemisphere is occurring at approximately 3 degrees of the constellation of Pisces, depending on which constellations division once wishes to use. There are three options.

A clear definition of the difference between these two things appears to have eluded Biodynamics, even though I believe, Dr Elizabeth Vreede, gave clear indications in her 'Astronomical Letters', in the 1930s.

Another lingering issue for Biodynamics is their commentators, often do not clearly defined the difference between Astronomy and Astrology. Astronomy is the study of the heavens, while Astrology is when any of those observations are then reflected back onto anything on Earth, to infer some 'meaning'. We are using the Constellations in planting calendars and the Signs as horoscopes for Humans. This topic and Biodynamics in general is therefore primarily an Astrological conversation.

Given this and that the Universe and Astrology are a vast subject, I have found it useful to define the **'context'** within which one is operating. What is the **reality** of the phenomena you are observing? I have found, **the 'astronomical' context of something, defines it meaning, and its applications**.

With this question we luckily have a firm reality to start from – **the Galaxy**. This is a real thing. However to get to US, the Humans, we need to take several steps 'AWAY', from this reality. Every time we take a step away, we change the context of the activity, and therefore we change its possible application.

We are going on a journey through the energetic onion of Creation. We are following one vibrational strand – the 12 foldness of the Constellations – but each time we change the context, it is as if we are moving one ring further inside the onion. We are going from the macrocosm – the Galaxy, to You and the exact spot you are on this planet. Quite a journey. One that takes us from the most extreme objective reality possible, to the most subjective reality possible, as our protagonists are indeed at either end of this spectrum.

We start with **real stars** and real constellations in the sky. These are stars beaming real electro magnetic forces at us, for billions of years. However we are seeing them, and **experiencing them from our position** in creation.

Step 1 - THE **primary reality**, The stars we see as constellations, are part of the Milky Way galaxy, and their reality is always in response to the **Galaxy,** based upon its centre, and its overall activity. In the Galaxy, the stars in the constellations we see, are not in the same relationship to each other, as we see them. This is our view only. These Star forces are called Spirit forces by Dr Steiner, and the Galaxy can be considered the local representative of the overall **Cosmic Spirit** realm, which includes all the rest that is out there. This real cosmic activity, is the energetic holographic ocean we float within, and it acts as a very real energetic base note, under everything we experience, but regardless of our view of it.

Step 2 - The first step 'away' we are making is to **see the constellations from our position**. So we now have a **subjective experience of the Cosmic Spirit**. We still experience the base note and now we receive it and have our experience of it. This is were the **Sidereal Constellational Zodiac** Dr Vreede described exists. Real stars , with our subjective experience of them. We experience those stars as formative clusters, beaming 'chords' of energy, that eventually manifest as the Species of Life.

The movement of the equinox through the constellations, is very widely appreciated to accurately describe the development of humanity, through its historical periods. Even the most hardened materialist historians appear to accept this cosmic influence upon our destinies. This 25750 year cycle described here is a very deep unconscious, unfolding of human evolution, imperceptible to

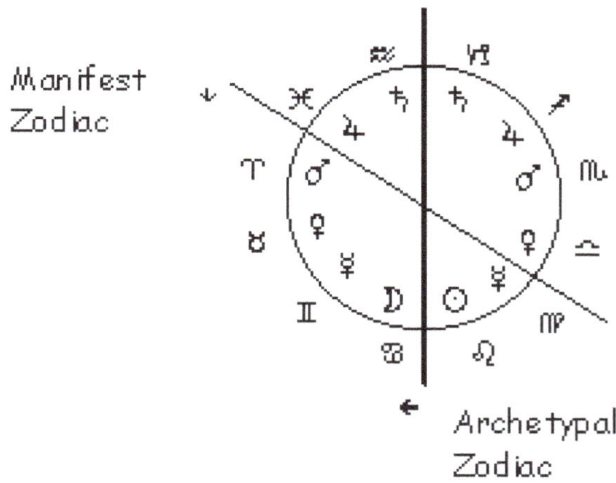

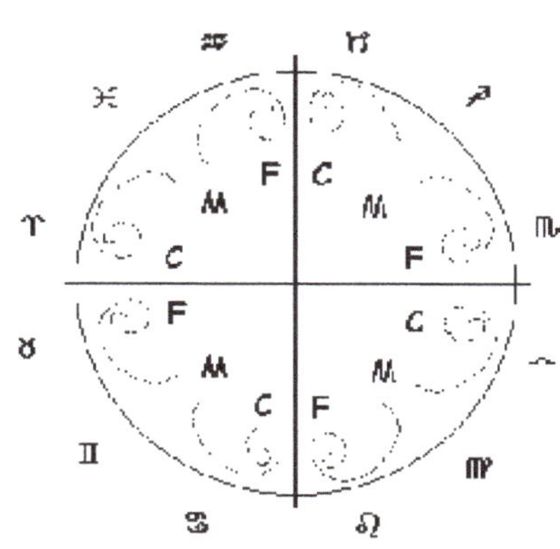

the average Human in their daily life. These are the great currents history beyond any one civilisation, let alone any individual. This is a fundamental rhythm.

We can then note that the Indian culture have long used the equal constellations, as the basis of their Vedic Astrology. Their focus is very fatalistic and focused upon the prediction of physical events in a persons life. There is no free will, there is no escape from the Astral prison. As the star prescribe, so it shall be.

Eugen Kolisko used them to outline the evolution of the animal kingdom, (19) while I showed how Chemistry organises according to the planets that rule the various constellations. (1) These are – archetypal organisations, that show a deep cosmic 'base tone' that effects evolution and the organisation of matter itself. These studies of the constellations also show, that they do not follow the Aries to Pisces track. The Equinox moves backwards through the Zodiac, and its starting point is Cancer. RS places the first post Atlantean period with Cancer, while the planetary rulers of the constellations, also point to Cancer as the Moon 1, beginning of the natural cycle with The Protozoa, running backwards, Gemini, Taurus etc and finishing with Leo as Moon 2 / Sun finale, The Mammals.

Step 3 - **Where are we?**

We are in a **Solar System**, which is only a very small speck, on one of the outer arms, within the Galaxy. It is based upon a Star, albeit quite a small one. This is our **World Spirit,** and from this context we draw **Heliocentric charts.** Willie Sucher (20) uses these charts to describe the deep spiritual intentions of individuals.

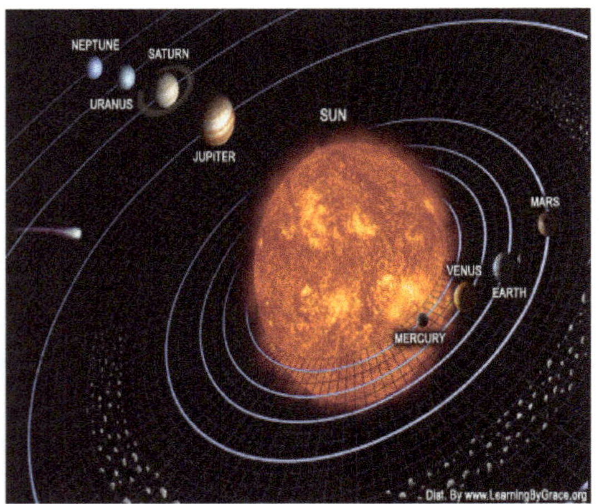

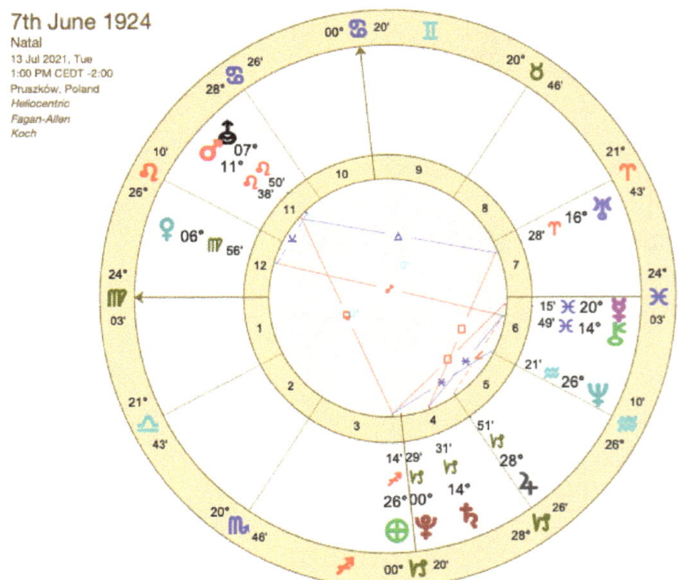

Step 4 - The next step is we are on a **planet** revolving around the Sun. This planet is a compost heap of substance, while the Sun is a nuclear reactor. Quite different, so we are moving away from World Spirit into the realm of the **World Astrality** – the planetary sphere. But also one step further towards physical matter. We are now on a planet.

Part of this move is when we look out from the Earth, we place the World Spirit Sun as part of the World Astrality. Dr Steiner uses the planets according to the length of their rhythms; Moon, Mercury, Venus, Sun, Mars, Jupiter, Saturn. These are real organic experiences for us.

In doing this though, we have subdued the **World Spirit as a possible 'captive' of the World Astrality**. We are empowering the World Astrality, which can easily overcome the World Spirit activity. This becomes mirrored the usual Human condition, of a Spirit, astrally possessed. **Geocentric charts** are made from this context. This is the **'What'** of our human journey. The planets tell us what the issues are. This still has constellations as it basis.

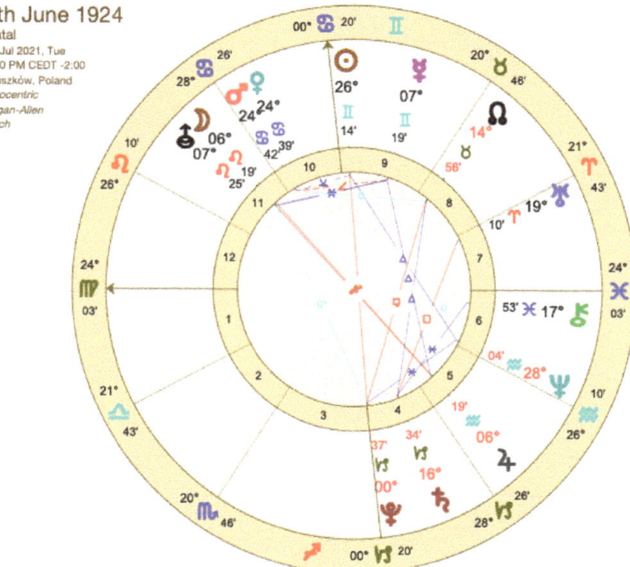

Step 5 – 'WE', The Lifeforms Throughout this story it has been us Humans who are looking out. Yet 'we' are a complex creation. We share this planet with the other 3 kingdoms of nature, and in Dr Steiner's stories of 'As Above , So Below' the kingdoms of nature arise by 'winding in' an extra external sphere. The Mineral kingdom just uses physical substance, while the plant kingdom, winds in the Earth's Atmospheric activities, we call an Etheric body, the Animals progressively wind in the Planetary sphere, to provide them with a Astral body, while we Humans have reached a stage where we are winding in the Star forces, we call Spirit. In doing this 'incarnation of the Spirit' activity we are taking the **World Spirit** activity of our Sun, and personalising it **into the Internal Spirit**. This is the part of us that allows for perceiving and discernment, leading to conscious decisions, and who is reading this. Hullo.

We need to reflect at this moment on Dr Steiner's evolution story, and how it was at the time of Christ, 0AD, that the Individual Spirit began to incarnate more deeply into the average Human. So here we have an element of time. It is now 2000 years since the Individual Spirit began separating us from the tight 'cosmic strings'. We began our 'liberation' from nature, so that today, at the end of the Age of Pisces, there are many humans who do not see themselves as being part of nature, in general. They need to exist in a completely artificial environment. We need to also note that it was this period 2000 years ago, that the 'Signs' of the zodiac were proposed by Ptolemy.

When the Signs were 'discovered' the Spirit begins to free Humans from the direct Cosmos.

From this point onwards in our journey, it is US **Humans making conscious choices**, about what we focus upon. We are defining the context. Each step made from here on is mirrored by **a step inside of us.**

Step 6 - The Signs of the Zodiac. This zodiac is not 'real'. It exists only in the minds of humans. It is **based upon the Suns path around the Earth**, and **starting from the spring equinox of the northern hemisphere**, upon which we mark off **12 equal divisions**.

So lets 'unpack' this

In step 4 we put the **Earth in the middle**, and have the **Sun moving around it**. In step 6 we are focusing our attention upon the **northern hemisphere and the spring equinox** of the Earth. The spring equinox is a real thing, and nature responds to it, and by choosing this Earth based phenomena we take another step towards the physical or 'personal' spheres.

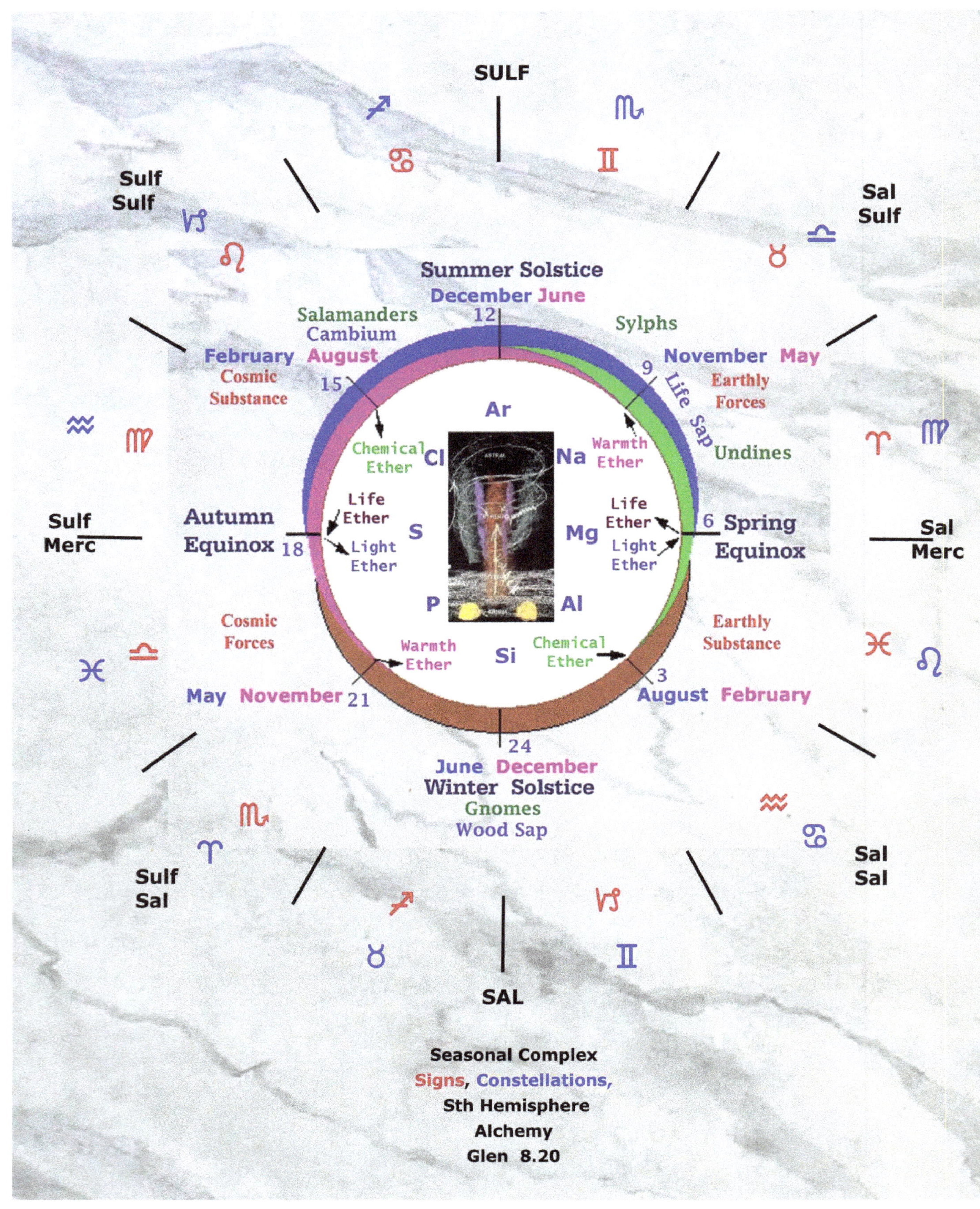

SULF
Sulf Sulf
Sal Sulf
Sulf Merc
Sal Merc
Sulf Sal
Sal Sal
SAL
Summer Solstice
December June
12
Salamanders
Cambium
February August
Cosmic Substance
15
Sylphs
November May
Earthly Forces
9
Life Sap
Ar
Chemical Ether
Cl
Na
Warmth Ether
Life Ether
Light Ether
S
Mg
Life Ether
Light Ether
Autumn Equinox
18
6 Spring Equinox
Undines
P
Al
Cosmic Forces
Warmth Ether
Si
Chemical Ether
Earthly Substance
3
May November
21
August February
June December
Winter Solstice
Gnomes
Wood Sap
24
Seasonal Complex
Signs, Constellations,
Sth Hemisphere
Alchemy
Glen 8.20

This is the **World Etheric** sphere. The Atmosphere within which the seasons occur and the living physical reality all lifeforms exist within.

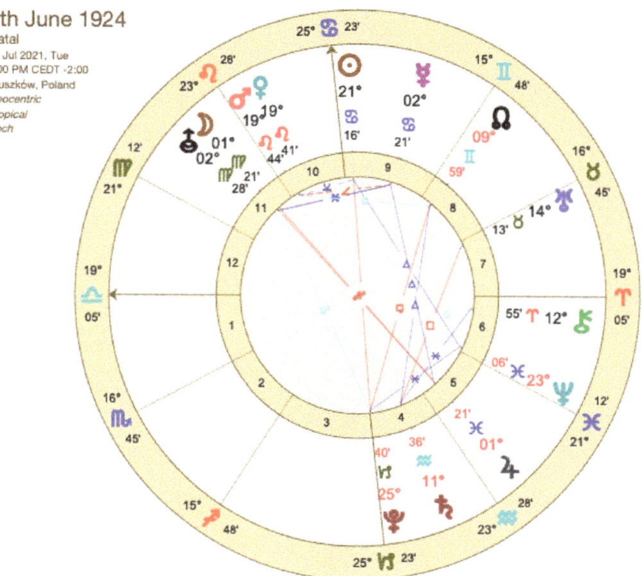

There is however one problem with the Spring Equinox, this is NOT the beginning of the seasonal year. Any gardener will tell you it begins at the winter solstice. Nature begins to move outward from mid winter, with a big leap forward being made about one month after mid winter, when the 'chemical ether' pushes bulbs to flower, shots to burst, and grubs to move upwards. The Spring Equinox is when we see all this activity burst fully above the soil. This is the time when the 'light ether' gains strength and the plant life generally begins to make harder stems and begin growing fruit. The 'Life Sap' processes take over. So **Spring Equinox** is a very specific position, and one quarter of the way through the annual seasonal cycle. It is **when things come to the surface and can be seen.**

 Even so, this is only real for half the planet, at a time. The southern hemisphere is experiencing something opposite to what is occurring at the northern hemisphere. We are making a choice. Initially because our culture developed in the northern hemisphere, but also due to the very real physical Earths magnetism and its predominant focus upon Magnetic North, no matter where you are on the planet. It does give this point some 'archetypal' relevance.   We are making a 'subjective and rational' choice, and thus another step towards the Human.

It is this consideration that led Hermann to suggest in his article, we might explore a Signal zodiac beginning at the Southern hemisphere spring equinox. This is not a new suggestion, however finding how this might have relevance for plant growth, other than seeing the seasons in a 12 fold division, is hard to determine. How can this be used in plant management. The real constellations will still be a strong influence on beings embedded in the Cosmic ocean.

Note we have moved a long way from the Cosmic Spirit and the constellations, which are Step 2.

Step 7 – 12 equal divisions. The third part of this statement is the equal 12 fold division of the Sun's path. There are 12 constellations generally and historically accepted within the Sidereal zodiac, however they are all of unequal size, so again we are making a human choice. We are taking the real 'archetypal' rhythm of the 12 constellations, and we are defining them to 'equality', and then we are placing this 'Cosmic Spirit' activity, onto the very physical northern hemisphere seasonal cycle. This is all a Human activity, we are setting these 'limits'. Thus more steps towards the personal.

In my studies of chemistry I found that the uses of the individual elements, provided a useful images of their inner qualities. So with the Signs, how are they used. Originally, 2000 years ago the human body parts were seen in relation to Signs. Aries governs the Head, Taurus the Throat and so on. These influences are still accepted today, as there is ample proof that Pisces often have difficulty with their feet, while Libras are very prone to kidney problems and so on. However as time has progressed, by far the greatest present use of the **Signs** of the Zodiac, are to describe **Human psychology**. The common 'horoscopes' in magazines describe indications of emotions and personal interactions and how you are going to feel about things. My mother, Judy, was a very competent Sun Sign astrologer. Indeed modern western Astrology has evolved Jungian inspired Humanistic

Astrology, where the 'inner life' of the Human is all they consider, and by consciously knowing this inner Astral dialogue, one can exert some free will, essentially **freeing the Spirit from its Astral prison.**

So in keeping with the 6 steps away from reality we have taken, once we arrive at the Signs, we have moved to this very abstract thing called **Human Psychology,** in general.

Step 8 – The Birth Chart, (18) This is based upon a persons Birth time, date and place. This final step to YOU sees these considerations find the exact place and time of a persons birth, within the cosmos. This personalises everything one step further, so that we can talk exactly about the individuals specific psychological issues and predict the events of their lives.

These exact details also allow for the placement of the **Houses**. So while we place the Signs at the Spring equinox, a similar action is taken by beginning the 12 Houses, from the Eastern Horizon at the time and place of Birth. This is called the Ascendant. Using this very physical reference, we are able to define the '**Where**' of a persons life. Where will their psychological emphasis' manifest physically throughout their lives. Work, Home or Politics etc. The Ascendant is seen as '**how we appear to others'**. This is our persona, our presentation, **what is seen first**. It is us at our most superficial, 'what comes to the surface first'. Just like at Spring.

The '**When',** of this story, is found by moving the planets in the sky, over the Birth positions.

All this movement from the Constellations to the Signs has occurred within us, because over the last 2000 years the individual Spirit has incarnated, breaking us free of the 'cosmic' reality, and bringing rational objective choice into play.

So a straight jump from the Constellations to the Signs, that people like to make, is not really valid. They do not start in the same place. Aries is not the first constellation, it is the fourth.

When looking at the planetary order and rulerships, one has to wonder why any zodiac would start at Aries.

Conclusion

This difference between the Constellations influence and the Signs is described in the journey we have taken. The closer to the cosmological reality we are, the more 'collective' and unconscious the influences will be. While the further we travel down the road of Human induced subjectivity, the more we move towards the internal reality of the Human.

Planets and animals do not have an incarnated Spirit, and so they do not walk the 'internal subjectivity' we Humans can, and so they remain embedded in the forces of the constellations.

The Signs of the zodiac are not directly relative to Biodynamics, they are a Human thing. Biodynamics rightly uses the constellations and all of the plant trials, I am aware of, carried out to look at celestial influence has found no relationship whatsoever to the Signs. Yes 'sloppy' people like to take Ms Thun's findings and place them over the Signs, however there is no proof of this effect. It is a commercial convenience by lazy people.

Through this journey I trust it is clear the Signs talk of very subjective human manifestation, that the rest of nature knows nothing about, while the Constellations transmit real formative, yet deeply unconscious vibrations through space, that fully influence the lower forms of Life. These are very different things.

This article was originally written for Harvest Magazine Vol 72-3. and was edited by Rebecca Reider

Constellations to Signs

1	Universe	Galaxy	Reality as it is	Spirit
2	Constellations	Star Groups	as we see the Galaxy	
3	Solar System	Sun	Heliocentric	
4	Planet	Earth	Geocentric	
5	Lifeforms	Us	the Individual	
6	Signs	Northern Hemisphere	Inside the Individual	
		Spring equinox	Physical Body order	
7	12 equal division		Psychology of the Individual	
8	Houses	Birth Place	Where Karma occurs	
				Manifestation

Essay Conclusion

Ok...that has been quite some journey. I appreciate this has been the 'long' version of the story, however given it has been 100 years, that this question has not been addressed adequately, I thought it worth the effort. It has also been an opportunity to place the Agriculture Lectures within the larger context of Dr Steiner's Nature lectures. A process that has bought quite some confusion for many people. I hope the order presented to you here brings some clarity to your journey with the unseen forces.

I am very thankful to Dr Bernard Lievegoed, for the tracks he left us to follow, to the entrance of the planets story. From there we have to find our own way, which I trust I have provided. I can but hope my offerings will settle this outstanding question. If only. Well I hope it helps settle it for you, and allows you to find useful remedies for the problems you face with nature.

References

(1) Glenological Chemistry   Glen Atkinson

(2) Nature of Substance   R Hauschka

(3) 1922 medical lectures   R Steiner

(4) The Working of the Planets and the Life Processes in Man and Earth. B. Lievegoed

(5) Alchemical Chemistry   Glen Atkinson

(6) Cosmic Workings in Earth and Man lecture 5,
  Man as Symphony lecture 7, R Steiner

(7) Energetic Activities   Glen Atkinson

(8) Critique of Enzo   Glen Atkinson
  http://garudabd.org/2019/03/27/in-response-to-enzo-nastati-intro-to-bd-in-9-meetings/

(9) RS Plant Growth Story   Glen Atkinson
  http://garudabd.org/dr-steiners-plant-story/

(10) Equisetum and Fungus   Glen Atkinson
  http://garudabd.org/wp-content/uploads/http://garudabd.org/sites/garudabd.org/files/Equisetum-v7_0.pdf

(11) Quantum Agriculture, pg 67-71 Hugh Lovel

(12) Man and Earth pg 89   K Konig

(13) The Etheric Forces Problem   G Atkinson
  https://garudabd.org/wp-content/uploads/EFF-v6-1.pdf

(14) Rudolf Steiner's birthchart
  https://garudabd.org/2019/08/09/jordon-peterson-and-rudolf-steiner/

(15) Harvest – the BDANZ magazine vol 72, No2

(16) Spiral Astrology   http://old.garudabd.org/books/book4.html

(17) Seasonal Complex http://garudabd.org/wp-content/uploads/Seasonal-Complex-Nth-copy.jpg

(18) What is the Birthchart.   http://old.garudabd.org/books/4_6.html

(19) Twelve Groups of Animals   E Kolisko

(20) Star Journals 1967   W Sucher

(21) The Organisation of Life   G Atkinson

(22) Bernie and Glen   G Atkinson

(23)The Cycle of the year   R Steiner 31.3.23

(24) Anthroposophical Approach to Medicine: Lecture III
  Lecture: S-5063: 27th October, 1922 | Stuttgart | GA0314